ATLAS OF CLASSICAL HISTORY

ATLAS OF
CLASSICAL HISTORY

Fifth edition

Michael Grant

New York
OXFORD UNIVERSITY PRESS
1994

First published in Great Britain by
The Orion Publishing Group Limited
5 Upper St. Martin's Lane, London WC2H 9EA

Published in the United States of America by
Oxford University Press, Inc.
200 Madison Avenue
New York, N.Y. 10016, U.S.A.

Library of Congress Cataloging-in-Publication Data
Grant, Michael, 1914–
Atlas of classical history / Michael Grant.—[New rev. ed.]
p. cm.
Covers the Near East, ancient Egypt, Greece, and Rome; shows the period between 1700 BC and 565 AD.
"First published in Great Britain by the Orion Publishing Group Limited ... London"—CIP t.p. verso.
Rev. ed. of: Ancient history atlas/Michael Grant. 1974.
Includes index.
ISBN 0–19–521074–3 (hardback)
ISBN 0–19–521078–6 (paperback)
1. Geography, Ancient—Maps. I. Grant, Michael, 1914–
Ancient history atlas. II. Title.
G1033.G65 1994 <G&M>
911'.3—dc20 93–48331
CIP
MAP

Printing (last digit): 9 8 7 6 5 4 3 2 1

Printed in Great Britain

Preface

This is an atlas of the classical world – the ancient Greek and Roman world, which needs to be understood if we are to understand the world of today. To say that such an atlas could ever be a substitute for a historical survey would be an exaggeration. Nevertheless, geography is such a vital, indeed predominant, factor in ancient history – and such a difficult factor because of all the changes of names[1] – that the whole course of events often seems to mean practically nothing without maps, and without a lot of them, carefully devised.

Older classical atlases, apart from a varying degree of emphasis on physical aspects, tended to concentrate on political themes, and it is true enough that these stand in great need of maps. But the present volume attempts to cast the net wider, and to introduce economic, cultural, religious and other topics as well. There are also a number of town plans.

Modern research in archaeology and other fields has shown that the classical world cannot be grasped without some appreciation of what went before it. I have consequently started this book with a number of maps illustrating the Mediterranean world during the second millennium BC, and particularly during the period from 1700 BC onwards, when the international scene had already assumed a well-defined and complex appearance; and the story is carried onwards to offer brief illustrations of the Old Testament. At the other end of the story, the traditional terminal date of the ancient world, the year AD 476 when the last western emperor ceased to reign, is again not a very meaningful landmark, so I have carried on the tale until the reign of Justinian in the following century.

It will clear enough what a very great deal is owed to the talent of Arthur Banks for transcribing the written and spoken world into cartographic form. I am also most grateful to Julian Shuckburgh and Benjamin Buchan for all the assistance they rendered on behalf of the publishers, and I want to thank Jane Dorner for assistance with the index and C. R. B. Elliott for help with an earlier revised edition. Finally, I have to acknowledge a substantial debt to existing classical atlases, German and English. And I must single out, for a special word of gratitude, the *Atlas of the Classical World* edited by A. A. M. van der Heyden and H. H. Scullard for Messrs Nelson, and *Westermanns Grosser Atlas zur Weltgechichte* (Westermann, Braunschweig). N. G. L. Hammond's *Atlas of the Greek and Roman World in Antiquity* (Noyes Press, Park Ridge) is now fundamental; so is Routledge's new classical atlas.

For this fifth edition I have added new maps on the changing frontier of the Roman Empire (maps 72 and 73), on the persecution of the Christians (map 86) and on the Roman Empire in its final years (maps 88 and 90).

1994 MICHAEL GRANT

[1] Modern names are given after the ancient in the Index.

List of Maps

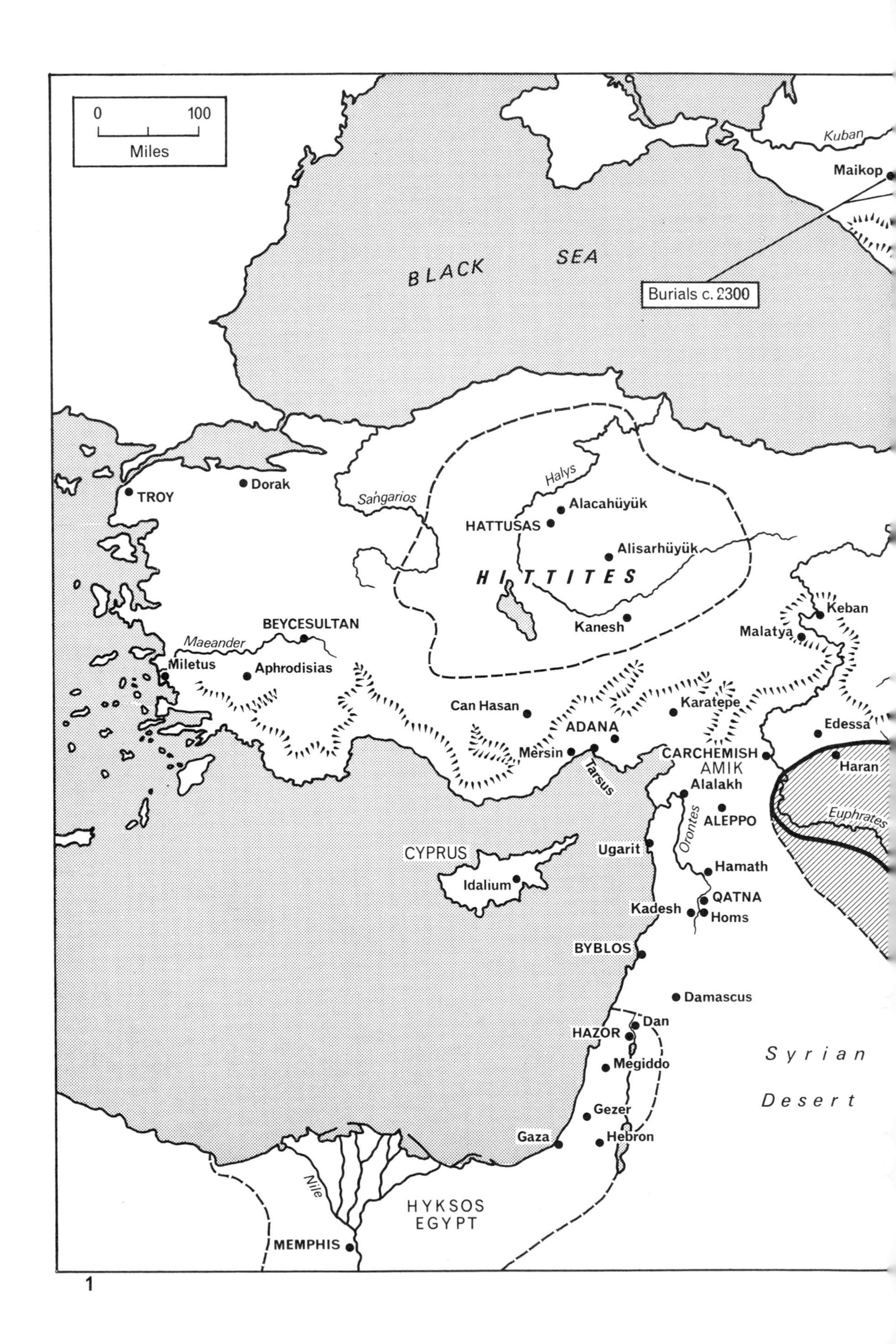
0
100
Miles
Kuban
Maikop
BLACK SEA
Burials c. 2300
TROY
Dorak
Halys
Sangarios
Alacahüyük
HATTUSAS
Alisarhüyük
HITTITES
Kanesh
Keban
Malatya
BEYCESULTAN
Maeander
Miletus
Aphrodisias
Can Hasan
Karatepe
ADANA
Edessa
Mersin
Tarsus
CARCHEMISH
AMIK
Haran
Alalakh
Orontes
ALEPPO
Euphrates
CYPRUS
Ugarit
Idalium
Hamath
QATNA
Kadesh
Homs
BYBLOS
Damascus
Dan
HAZOR
Megiddo
Syrian Desert
Gezer
Gaza
Hebron
Nile
HYKSOS EGYPT
MEMPHIS

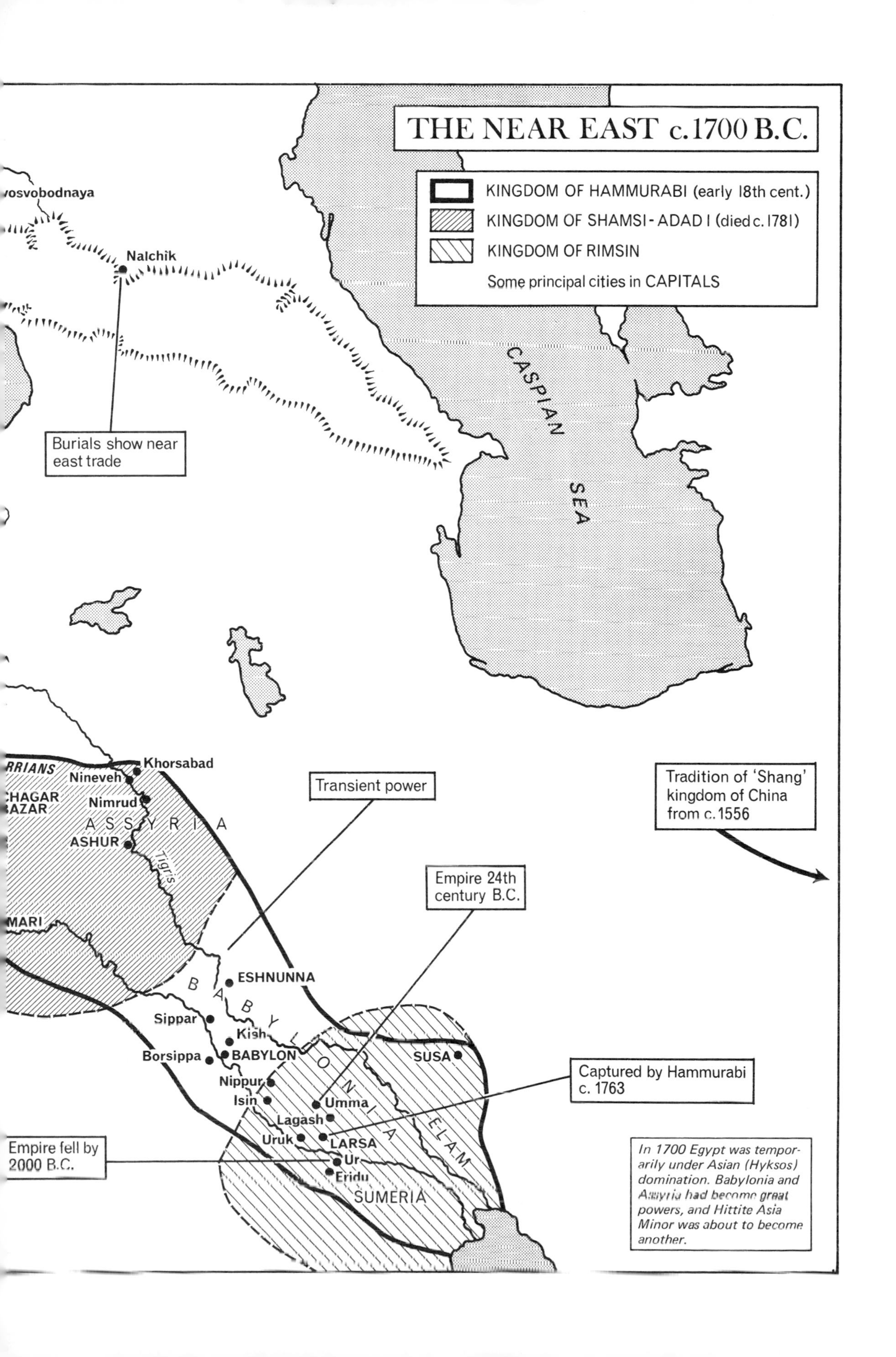
THE NEAR EAST c.1700 B.C.
KINGDOM OF HAMMURABI (early 18th cent.)
KINGDOM OF SHAMSI-ADAD I (died c.1781)
KINGDOM OF RIMSIN
Some principal cities in CAPITALS
osvobodnaya
Nalchik
Burials show near east trade
CASPIAN SEA
RRIANS
Khorsabad
Nineveh
HAGAR
AZAR
Nimrud
ASSYRIA
ASHUR
Tigris
MARI
Transient power
Empire 24th century B.C.
ESHNUNNA
BABYLONIA
Sippar
Kish
Borsippa
BABYLON
SUSA
Nippur
Isin
Umma
Lagash
Uruk
LARSA
Ur
Eridu
SUMERIA
ELAM
Captured by Hammurabi c. 1763
Empire fell by 2000 B.C.
Tradition of 'Shang' kingdom of China from c.1556
In 1700 Egypt was temporarily under Asian (Hyksos) domination. Babylonia and Assyria had become great powers, and Hittite Asia Minor was about to become another.

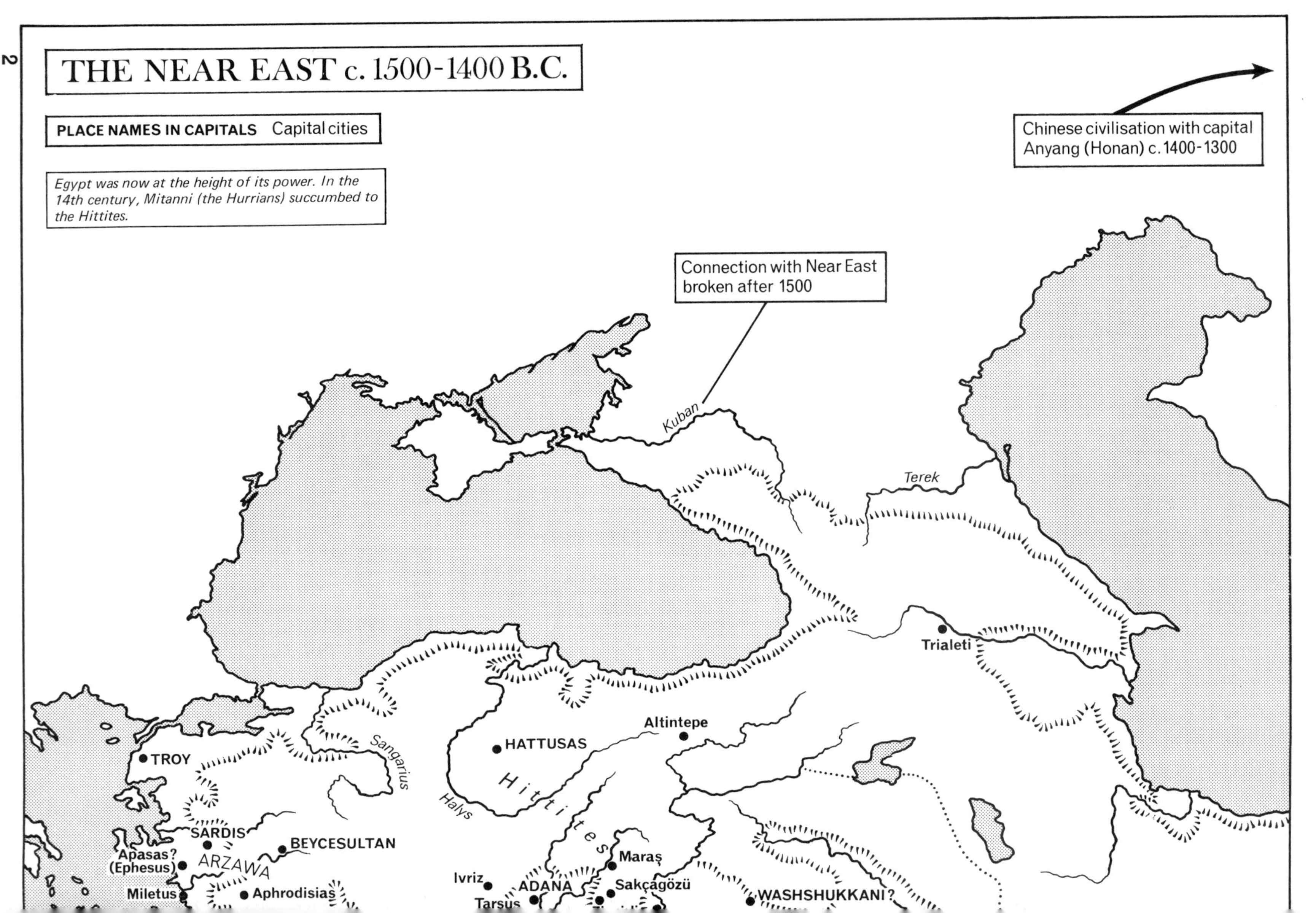
THE NEAR EAST c. 1500-1400 B.C.
PLACE NAMES IN CAPITALS Capital cities
Egypt was now at the height of its power. In the 14th century, Mitanni (the Hurrians) succumbed to the Hittites.
Chinese civilisation with capital Anyang (Honan) c. 1400-1300
Connection with Near East broken after 1500
Kuban
Terek
Trialeti
Altintepe
HATTUSAS
Hittites
Halys
Sangarius
TROY
SARDIS
Apasas? (Ephesus)
ARZAWA
BEYCESULTAN
Miletus
Aphrodisias
Ivriz
ADANA
Tarsus
Maraş
Sakçagözü
WASHSHUKKANI?

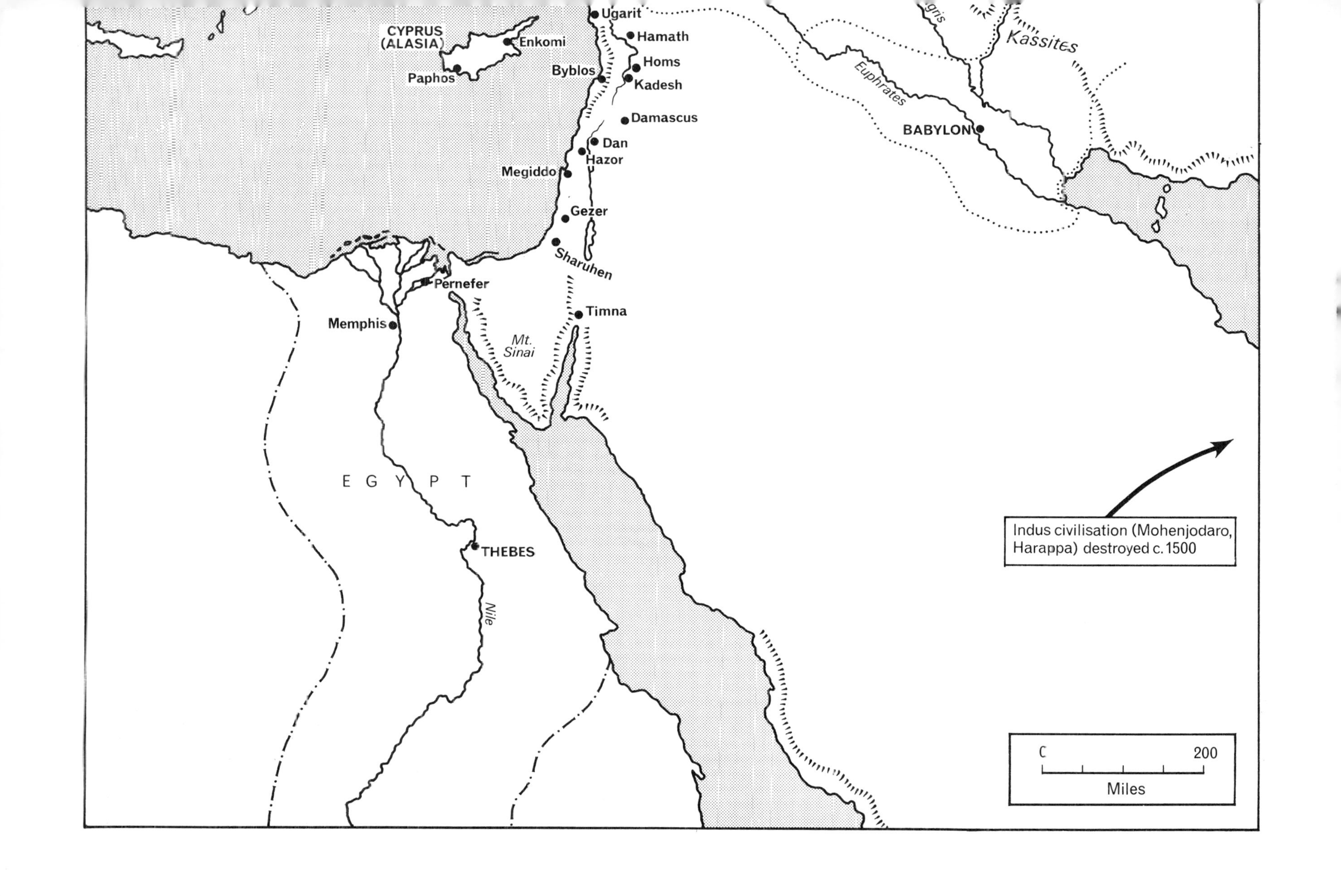
CYPRUS
(ALASIA)
Enkomi
Paphos
Ugarit
Hamath
Homs
Byblos
Kadesh
Damascus
Dan
Hazor
Megiddo
Gezer
Sharuhen
Pernefer
Timna
Memphis
Mt.
Sinai
EGYPT
THEBES
Nile
Euphrates
BABYLON
Kassites
Indus civilisation (Mohenjodaro, Harappa) destroyed c. 1500
C
200
Miles

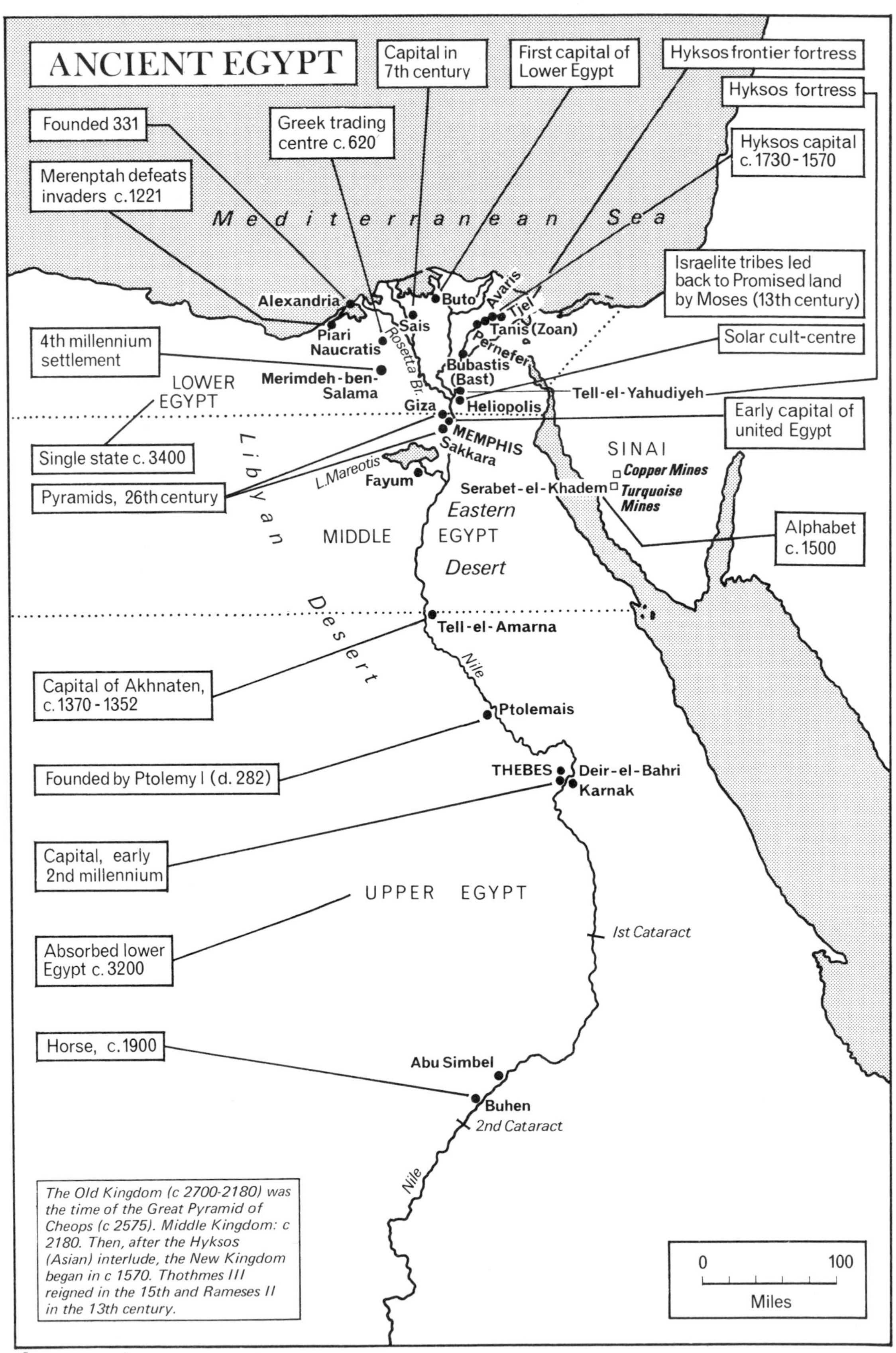
ANCIENT EGYPT
Capital in 7th century
First capital of Lower Egypt
Hyksos frontier fortress
Hyksos fortress
Founded 331
Greek trading centre c. 620
Hyksos capital c. 1730 - 1570
Merenptah defeats invaders c. 1221
Mediterranean Sea
Israelite tribes led back to Promised land by Moses (13th century)
Alexandria
Buto
Avaris
Tjel
Sais
Tanis (Zoan)
Piari
Naucratis
Rosetta Br.
Pernefer
Solar cult-centre
4th millennium settlement
Merimdeh-ben-Salama
Bubastis (Bast)
LOWER EGYPT
Tell-el-Yahudiyeh
Giza
Heliopolis
Early capital of united Egypt
MEMPHIS
Sakkara
SINAI
Single state c. 3400
Libyan Desert
L. Mareotis
Fayum
Copper Mines
Pyramids, 26th century
Serabet-el-Khadem
Turquoise Mines
Eastern Desert
MIDDLE EGYPT
Alphabet c. 1500
Tell-el-Amarna
Nile
Capital of Akhnaten, c. 1370 - 1352
Ptolemais
Founded by Ptolemy I (d. 282)
THEBES
Deir-el-Bahri
Karnak
Capital, early 2nd millennium
UPPER EGYPT
Ist Cataract
Absorbed lower Egypt c. 3200
Horse, c. 1900
Abu Simbel
Buhen
2nd Cataract
Nile
The Old Kingdom (c 2700-2180) was the time of the Great Pyramid of Cheops (c 2575). Middle Kingdom: c 2180. Then, after the Hyksos (Asian) interlude, the New Kingdom began in c 1570. Thothmes III reigned in the 15th and Rameses II in the 13th century.
0
100
Miles

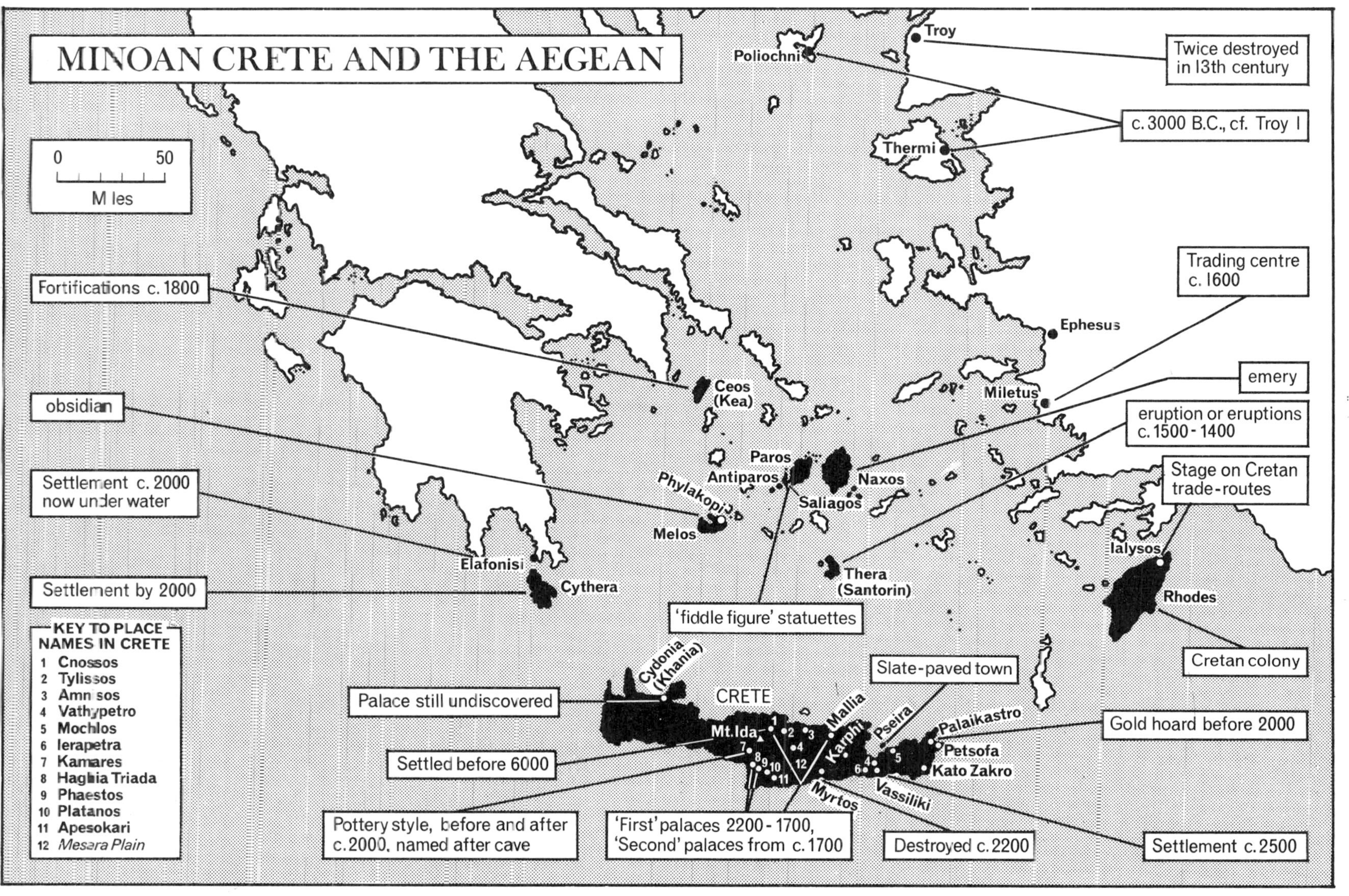

MINOAN CRETE AND THE AEGEAN
0
50
Miles
Troy
Twice destroyed in 13th century
Poliochni
c. 3000 B.C., cf. Troy I
Thermi
Fortifications c. 1800
Trading centre c. 1600
Ephesus
Ceos (Kea)
emery
Miletus
obsidian
eruption or eruptions c. 1500 - 1400
Paros
Antiparos
Naxos
Saliagos
Phylakopi
Melos
Settlement c. 2000 now under water
Stage on Cretan trade-routes
Elafonisi
Cythera
Settlement by 2000
Thera (Santorin)
Ialysos
Rhodes
'fiddle figure' statuettes
KEY TO PLACE NAMES IN CRETE
1 Cnossos
2 Tylissos
3 Amnisos
4 Vathypetro
5 Mochlos
6 Ierapetra
7 Kamares
8 Haghia Triada
9 Phaestos
10 Platanos
11 Apesokari
12 Mesara Plain
Cydonia (Khania)
CRETE
Palace still undiscovered
Mt. Ida
Mallia
Karphi
Pseira
Slate-paved town
Palaikastro
Petsofa
Kato Zakro
Gold hoard before 2000
Cretan colony
Settled before 6000
Myrtos
Vassiliki
Pottery style, before and after c. 2000, named after cave
'First' palaces 2200 - 1700, 'Second' palaces from c. 1700
Destroyed c. 2200
Settlement c. 2500

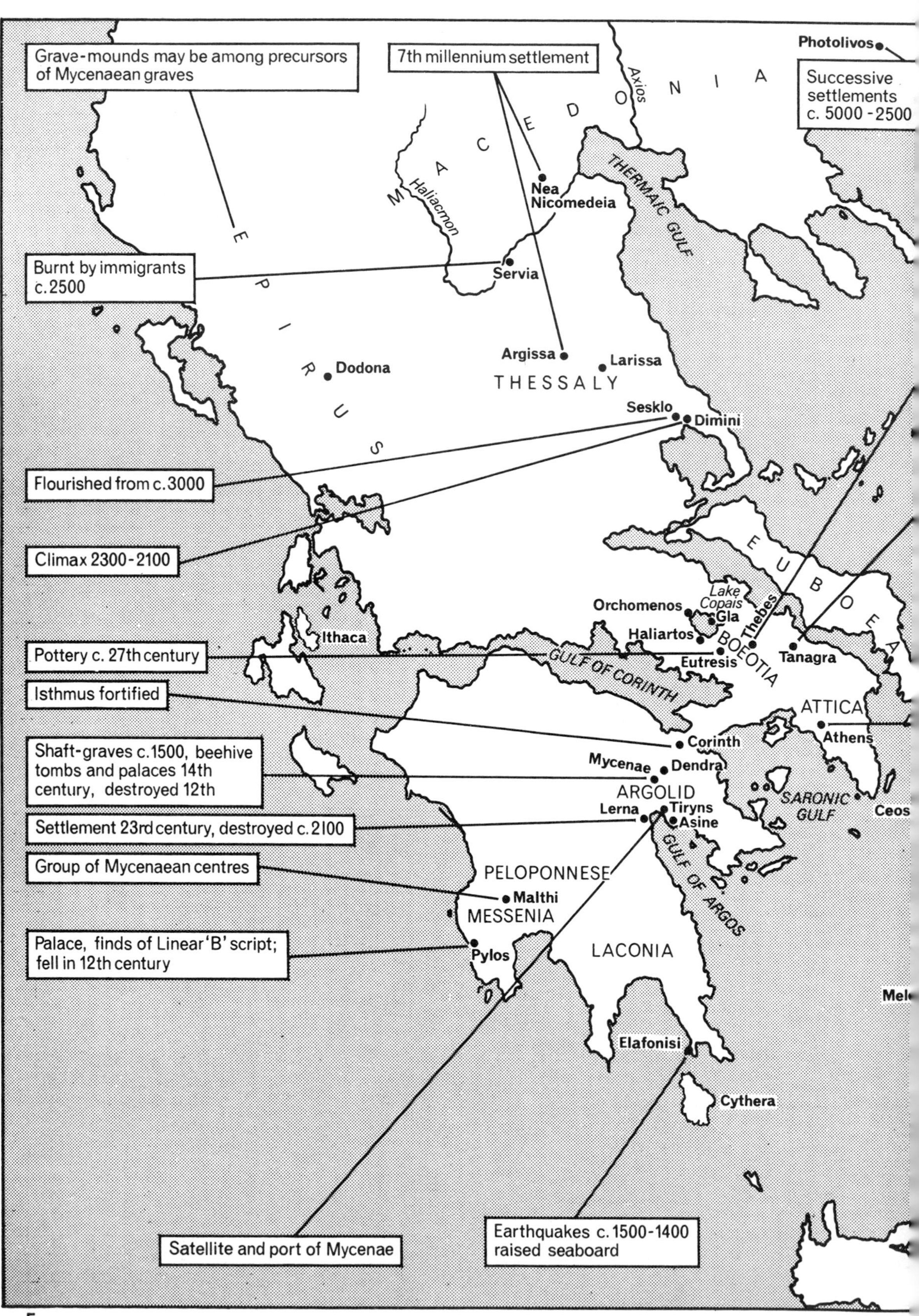
Grave-mounds may be among precursors of Mycenaean graves
7th millennium settlement
Photolivos
Successive settlements c. 5000 - 2500
Axios
M A C E D O N I A
THERMAIC GULF
Haliacmon
Nea Nicomedeia
E P I R U S
Burnt by immigrants c. 2500
Servia
Argissa
Larissa
Dodona
THESSALY
Sesklo
Dimini
Flourished from c. 3000
Climax 2300-2100
E U B O E A
Lake Copais
Orchomenos
Gla
Thebes
Haliartos
Ithaca
Pottery c. 27th century
BOEOTIA
Eutresis
Tanagra
GULF OF CORINTH
Isthmus fortified
ATTICA
Corinth
Athens
Shaft-graves c. 1500, beehive tombs and palaces 14th century, destroyed 12th
Mycenae
Dendra
ARGOLID
Lerna
Tiryns
Asine
SARONIC GULF
Ceos
Settlement 23rd century, destroyed c. 2100
GULF OF ARGOS
Group of Mycenaean centres
PELOPONNESE
Malthi
MESSENIA
Palace, finds of Linear 'B' script; fell in 12th century
Pylos
LACONIA
Elafonisi
Cythera
Satellite and port of Mycenae
Earthquakes c. 1500-1400 raised seaboard

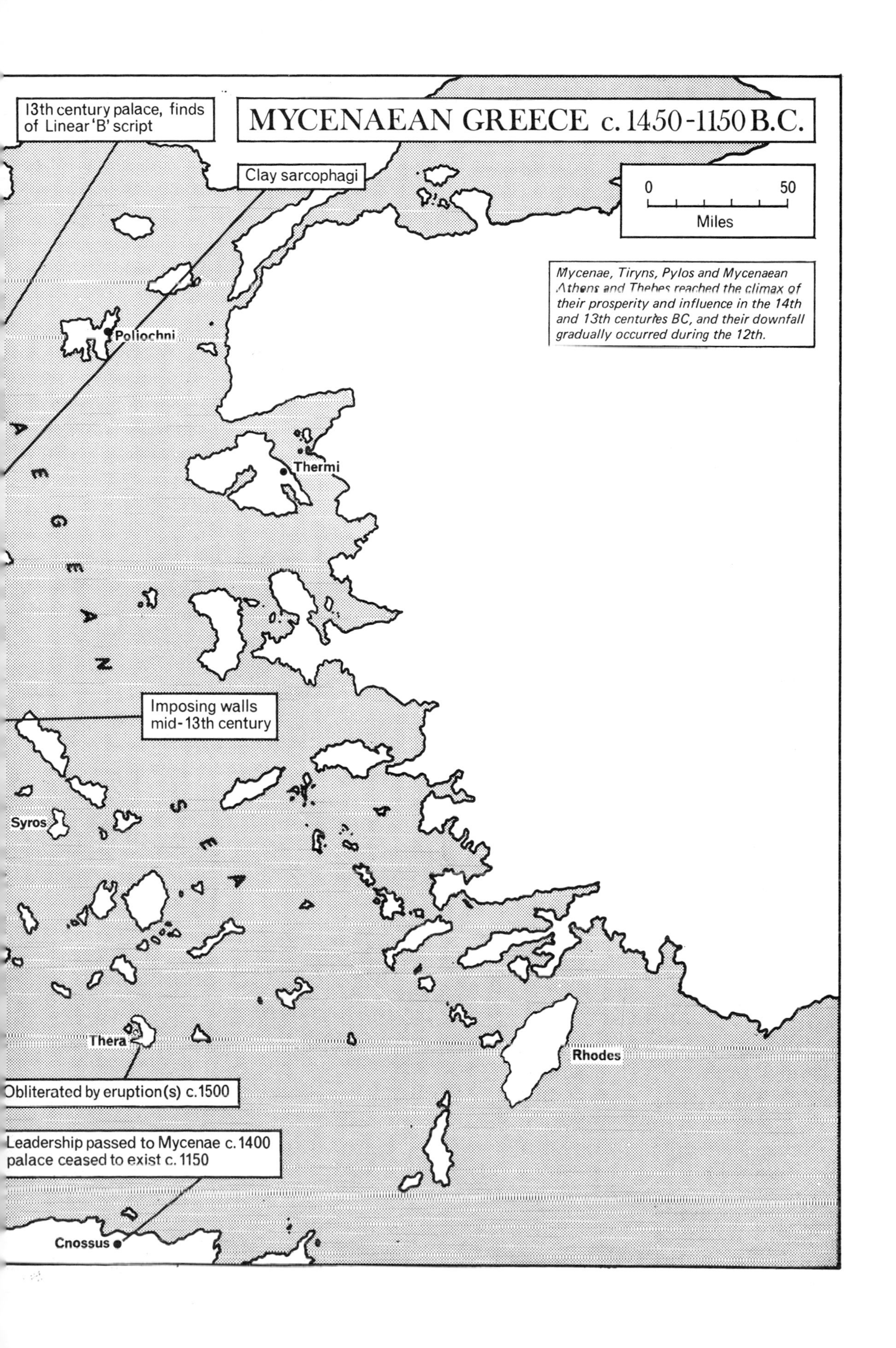
13th century palace, finds of Linear 'B' script
MYCENAEAN GREECE c. 1450-1150 B.C.
Clay sarcophagi
0
50
Miles
Mycenae, Tiryns, Pylos and Mycenaean Athens and Thebes reached the climax of their prosperity and influence in the 14th and 13th centuries BC, and their downfall gradually occurred during the 12th.
Poliochni
Thermi
A
E
G
E
A
N
Imposing walls mid-13th century
Syros
S
E
A
Thera
Rhodes
Obliterated by eruption(s) c.1500
Leadership passed to Mycenae c.1400 palace ceased to exist c.1150
Cnossus

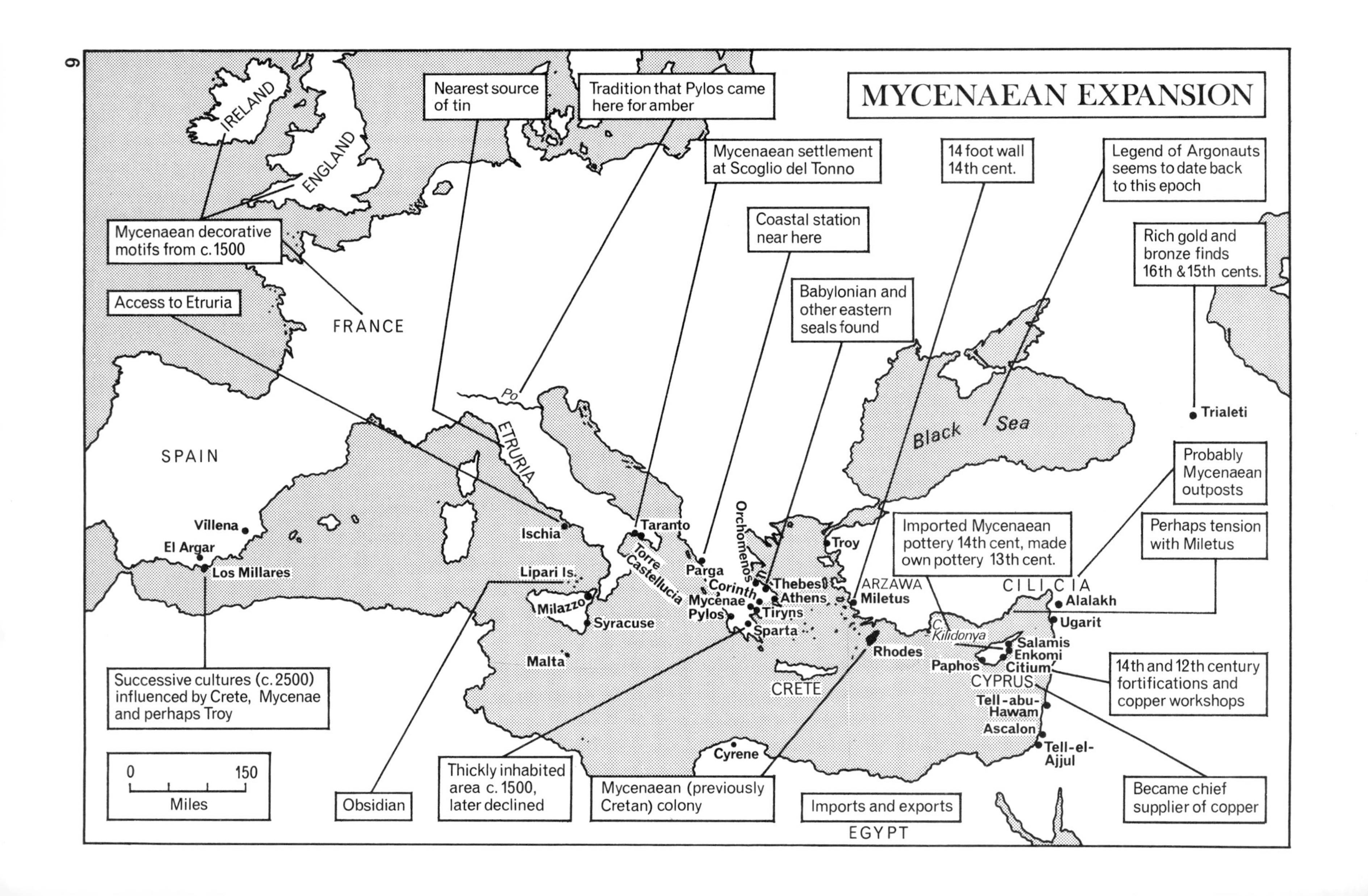
MYCENAEAN EXPANSION
Nearest source of tin
Tradition that Pylos came here for amber
Mycenaean settlement at Scoglio del Tonno
14 foot wall 14th cent.
Legend of Argonauts seems to date back to this epoch
Coastal station near here
Rich gold and bronze finds 16th & 15th cents.
Babylonian and other eastern seals found
Mycenaean decorative motifs from c. 1500
Access to Etruria
IRELAND
ENGLAND
FRANCE
SPAIN
Po
ETRURIA
Black Sea
Trialeti
Probably Mycenaean outposts
Perhaps tension with Miletus
Imported Mycenaean pottery 14th cent, made own pottery 13th cent.
Villena
El Argar
Los Millares
Ischia
Lipari Is.
Milazzo
Syracuse
Malta
Taranto
Torre Castellucia
Parga
Orchomenos
Corinth
Mycenae
Pylos
Thebes
Athens
Tiryns
Sparta
Troy
ARZAWA
Miletus
Rhodes
CRETE
CILICIA
Alalakh
Ugarit
C. Kilidonya
Salamis
Enkomi
Citium
Paphos
CYPRUS
Tell-abu-Hawam
Ascalon
Tell-el-Ajjul
Cyrene
14th and 12th century fortifications and copper workshops
Successive cultures (c. 2500) influenced by Crete, Mycenae and perhaps Troy
0 150 Miles
Obsidian
Thickly inhabited area c. 1500, later declined
Mycenaean (previously Cretan) colony
Imports and exports
EGYPT
Became chief supplier of copper

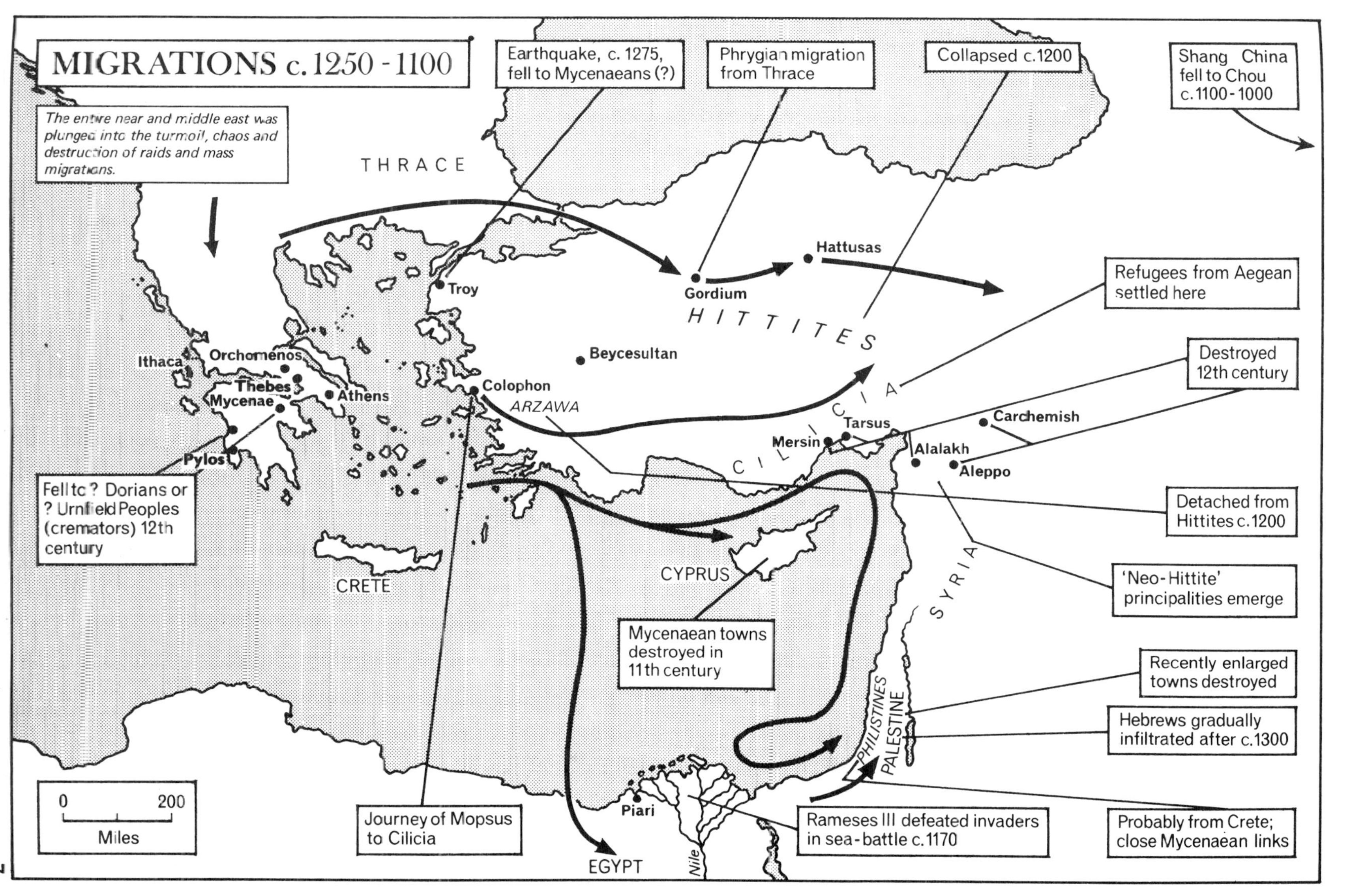
MIGRATIONS c. 1250 - 1100
The entire near and middle east was plunged into the turmoil, chaos and destruction of raids and mass migrations.
Earthquake, c. 1275, fell to Mycenaeans (?)
Phrygian migration from Thrace
Collapsed c.1200
Shang China fell to Chou c.1100-1000
THRACE
Troy
Gordium
Hattusas
HITTITES
Refugees from Aegean settled here
Destroyed 12th century
Beycesultan
Colophon
ARZAWA
Ithaca
Orchomenos
Thebes
Mycenae
Athens
Pylos
Fell to ? Dorians or ? Urnfield Peoples (cremators) 12th century
CILICIA
Mersin
Tarsus
Carchemish
Alalakh
Aleppo
Detached from Hittites c.1200
SYRIA
'Neo-Hittite' principalities emerge
CRETE
CYPRUS
Mycenaean towns destroyed in 11th century
Recently enlarged towns destroyed
PHILISTINES
PALESTINE
Hebrews gradually infiltrated after c.1300
0
200
Miles
Journey of Mopsus to Cilicia
Piari
EGYPT
Nile
Rameses III defeated invaders in sea-battle c.1170
Probably from Crete; close Mycenaean links

PHOENICIAN TRADE & COLONISATION

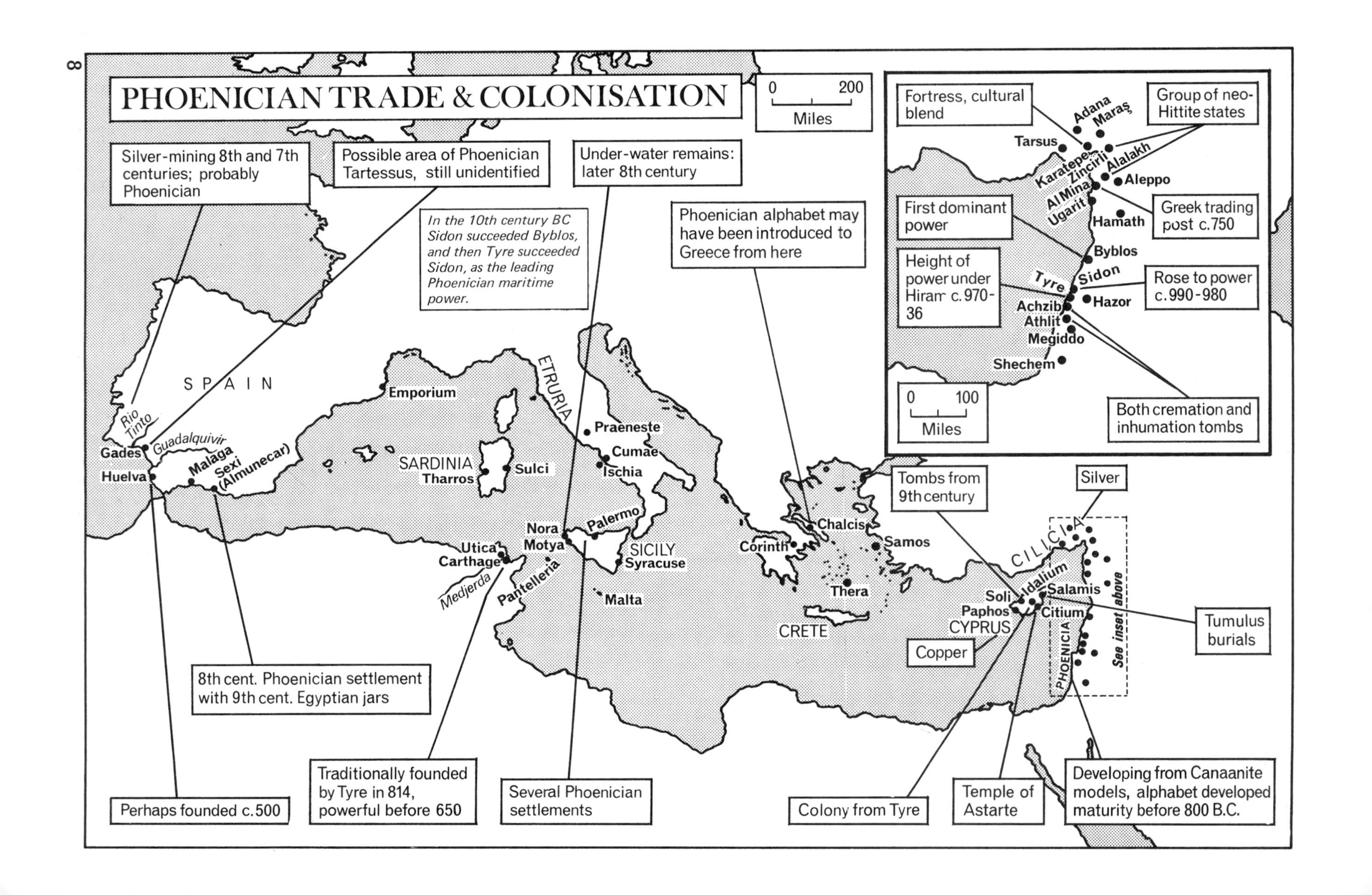

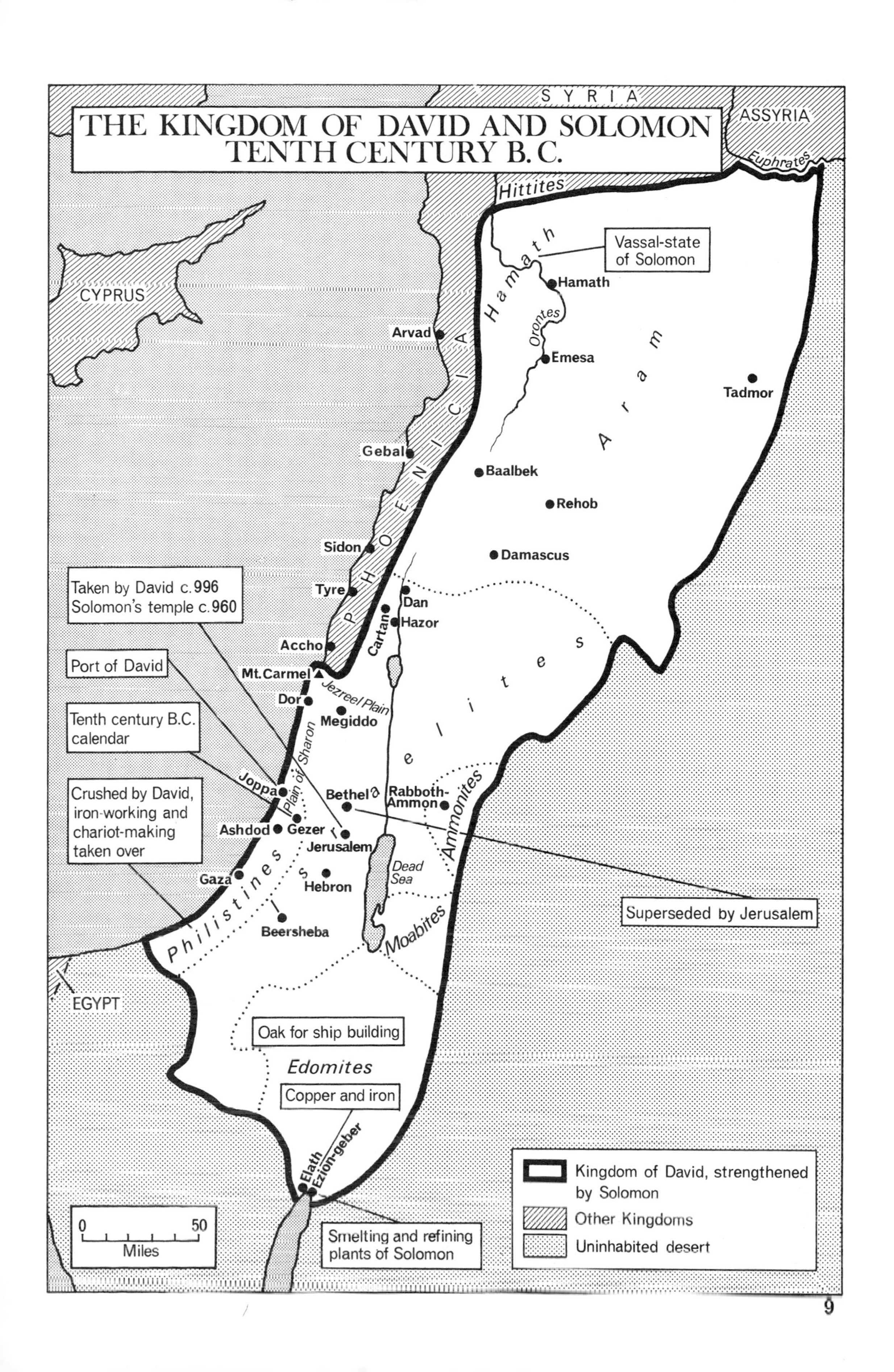

THE KINGDOM OF DAVID AND SOLOMON
TENTH CENTURY B.C.
SYRIA
ASSYRIA
Euphrates
Hittites
Vassal-state
of Solomon
Hamath
Hamath
Orontes
CYPRUS
Arvad
PHOENICIA
Emesa
Aram
Tadmor
Gebal
Baalbek
Rehob
Sidon
Damascus
Taken by David c.996
Solomon's temple c.960
Tyre
Dan
Cartan
Hazor
Accho
Israelites
Port of David
Mt.Carmel
Jezreel Plain
Dor
Megiddo
Tenth century B.C.
calendar
Plain of Sharon
Joppa
Bethel
Rabboth-Ammon
Ammonites
Crushed by David,
iron-working and
chariot-making
taken over
Ashdod
Gezer
Jerusalem
Dead Sea
Gaza
Hebron
Superseded by Jerusalem
Philistines
Beersheba
Moabites
EGYPT
Oak for ship building
Edomites
Copper and iron
Elath
Ezion-geber
Kingdom of David, strengthened by Solomon
Other Kingdoms
Uninhabited desert
0
50
Miles
Smelting and refining
plants of Solomon

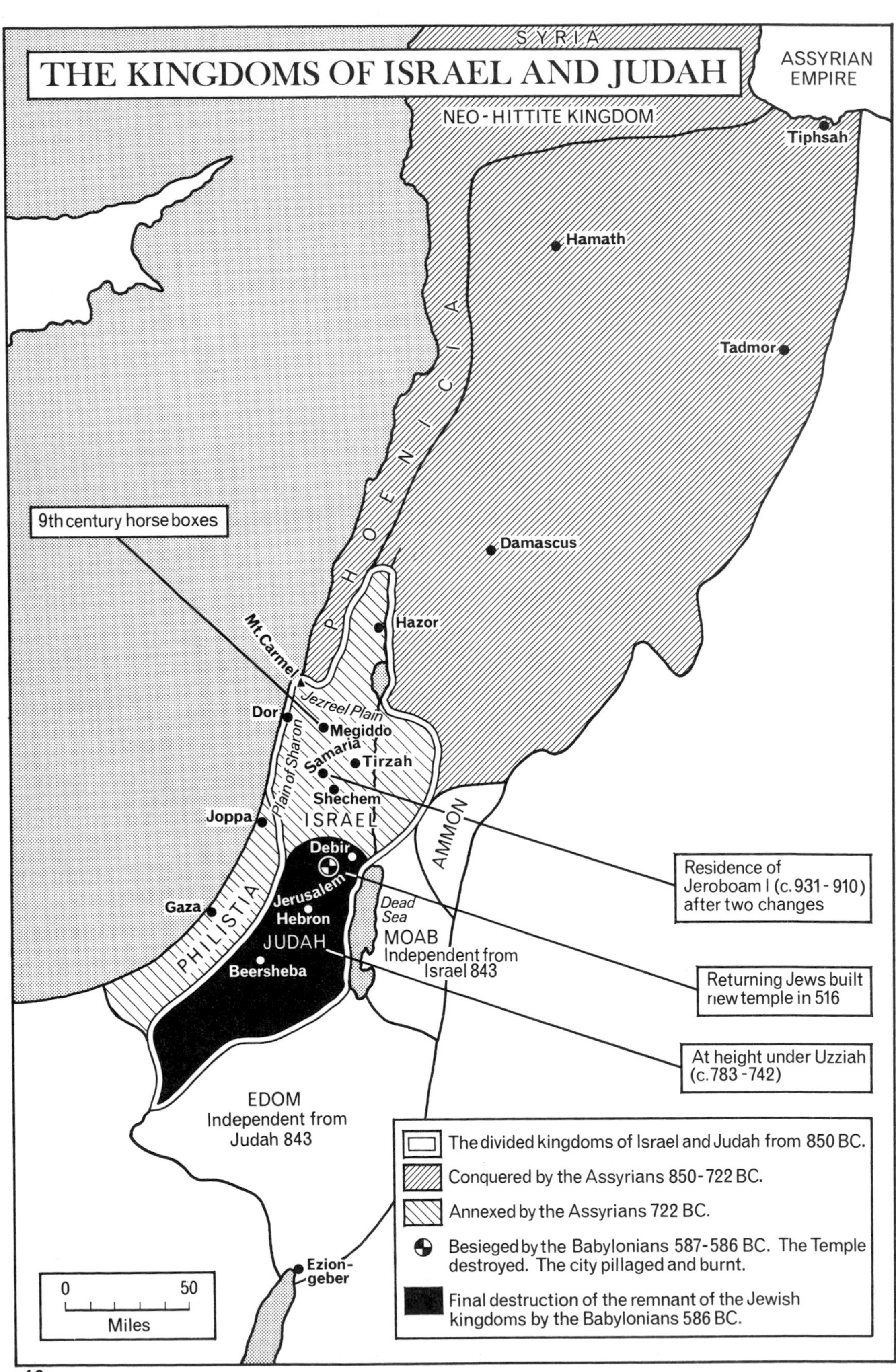
THE KINGDOMS OF ISRAEL AND JUDAH
SYRIA
ASSYRIAN EMPIRE
NEO-HITTITE KINGDOM
Tiphsah
Hamath
Tadmor
PHOENICIA
Damascus
9th century horse boxes
Hazor
Mt.Carmel
Jezreel Plain
Dor
Megiddo
Samaria
Tirzah
Plain of Sharon
Shechem
Joppa
ISRAEL
AMMON
Debir
Jerusalem
Hebron
Gaza
PHILISTIA
Dead Sea
JUDAH
Beersheba
MOAB
Independent from Israel 843
Residence of Jeroboam I (c.931-910) after two changes
Returning Jews built new temple in 516
At height under Uzziah (c.783-742)
EDOM
Independent from Judah 843
Ezion-geber
0
50
Miles
The divided kingdoms of Israel and Judah from 850 BC.
Conquered by the Assyrians 850-722 BC.
Annexed by the Assyrians 722 BC.
Besieged by the Babylonians 587-586 BC. The Temple destroyed. The city pillaged and burnt.
Final destruction of the remnant of the Jewish kingdoms by the Babylonians 586 BC.

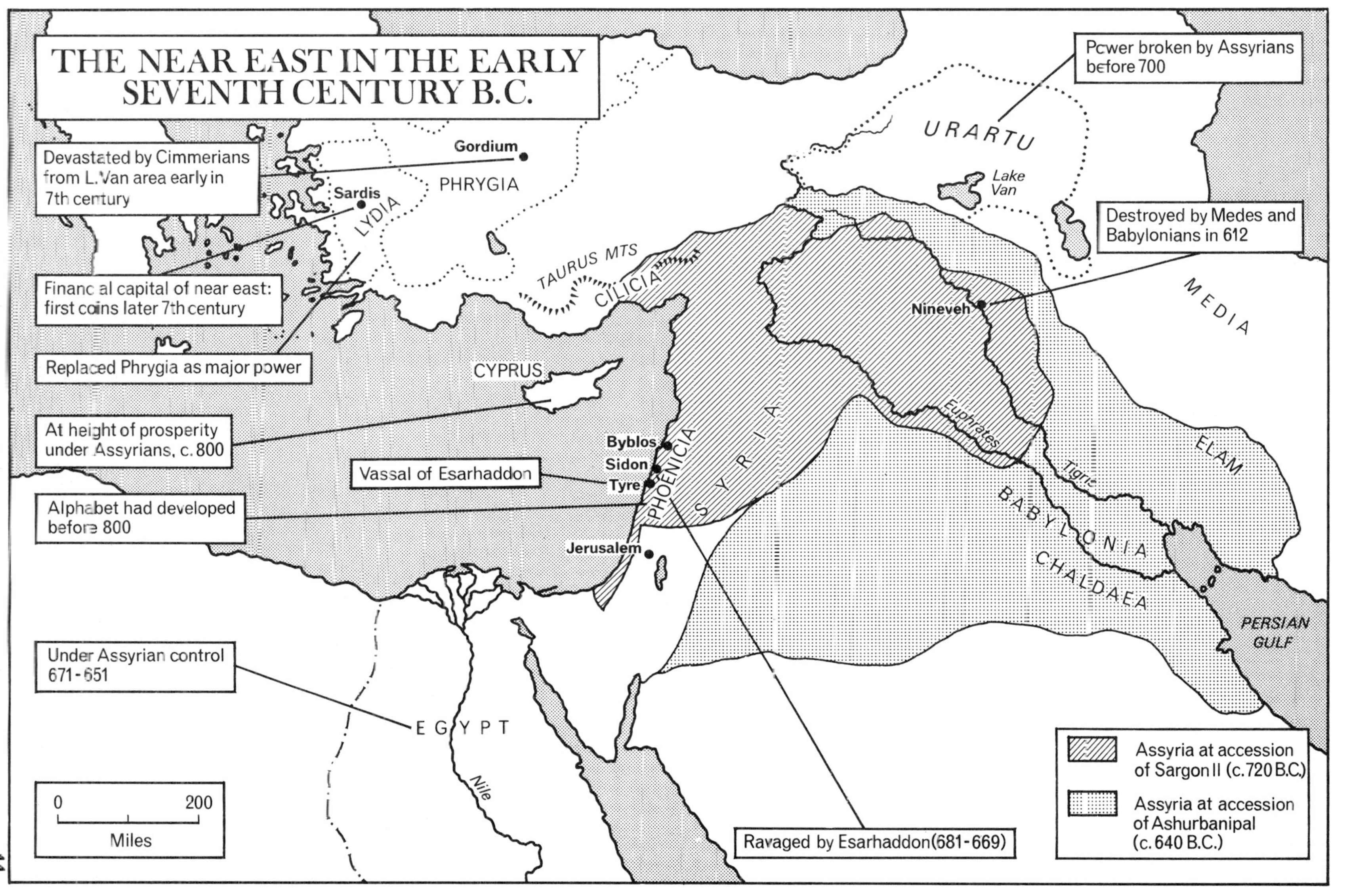

THE NEAR EAST IN THE EARLY SEVENTH CENTURY B.C.
Devastated by Cimmerians from L. Van area early in 7th century
Financial capital of near east: first coins later 7th century
Replaced Phrygia as major power
At height of prosperity under Assyrians, c. 800
Vassal of Esarhaddon
Alphabet had developed before 800
Under Assyrian control 671-651
Power broken by Assyrians before 700
Destroyed by Medes and Babylonians in 612
Ravaged by Esarhaddon (681-669)
Assyria at accession of Sargon II (c. 720 B.C.)
Assyria at accession of Ashurbanipal (c. 640 B.C.)
0
200
Miles
Gordium
PHRYGIA
Sardis
LYDIA
TAURUS MTS
CILICIA
CYPRUS
Byblos
Sidon
Tyre
Jerusalem
PHOENICIA
SYRIA
URARTU
Lake Van
MEDIA
Nineveh
Euphrates
Tigris
ELAM
BABYLONIA
CHALDAEA
PERSIAN GULF
EGYPT
Nile

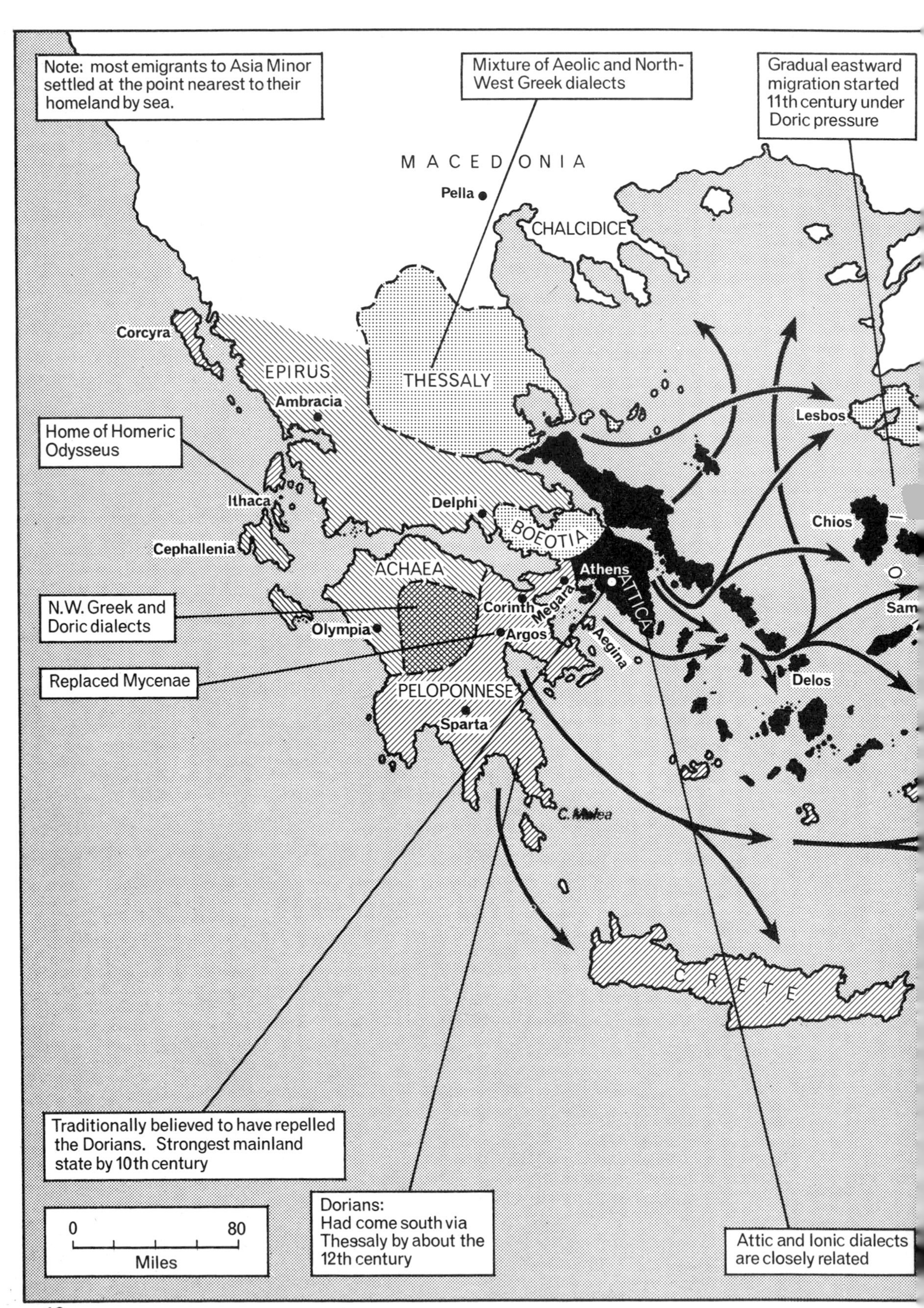

Note: most emigrants to Asia Minor settled at the point nearest to their homeland by sea.
Mixture of Aeolic and North-West Greek dialects
Gradual eastward migration started 11th century under Doric pressure
MACEDONIA
Pella
CHALCIDICE
Corcyra
EPIRUS
THESSALY
Ambracia
Lesbos
Home of Homeric Odysseus
Ithaca
Delphi
Chios
Cephallenia
BOEOTIA
ACHAEA
Athens
ATTICA
N.W. Greek and Doric dialects
Corinth
Megara
Olympia
Argos
Aegina
Replaced Mycenae
Delos
PELOPONNESE
Sparta
C. Malea
CRETE
Traditionally believed to have repelled the Dorians. Strongest mainland state by 10th century
0
80
Miles
Dorians: Had come south via Thessaly by about the 12th century
Attic and Ionic dialects are closely related

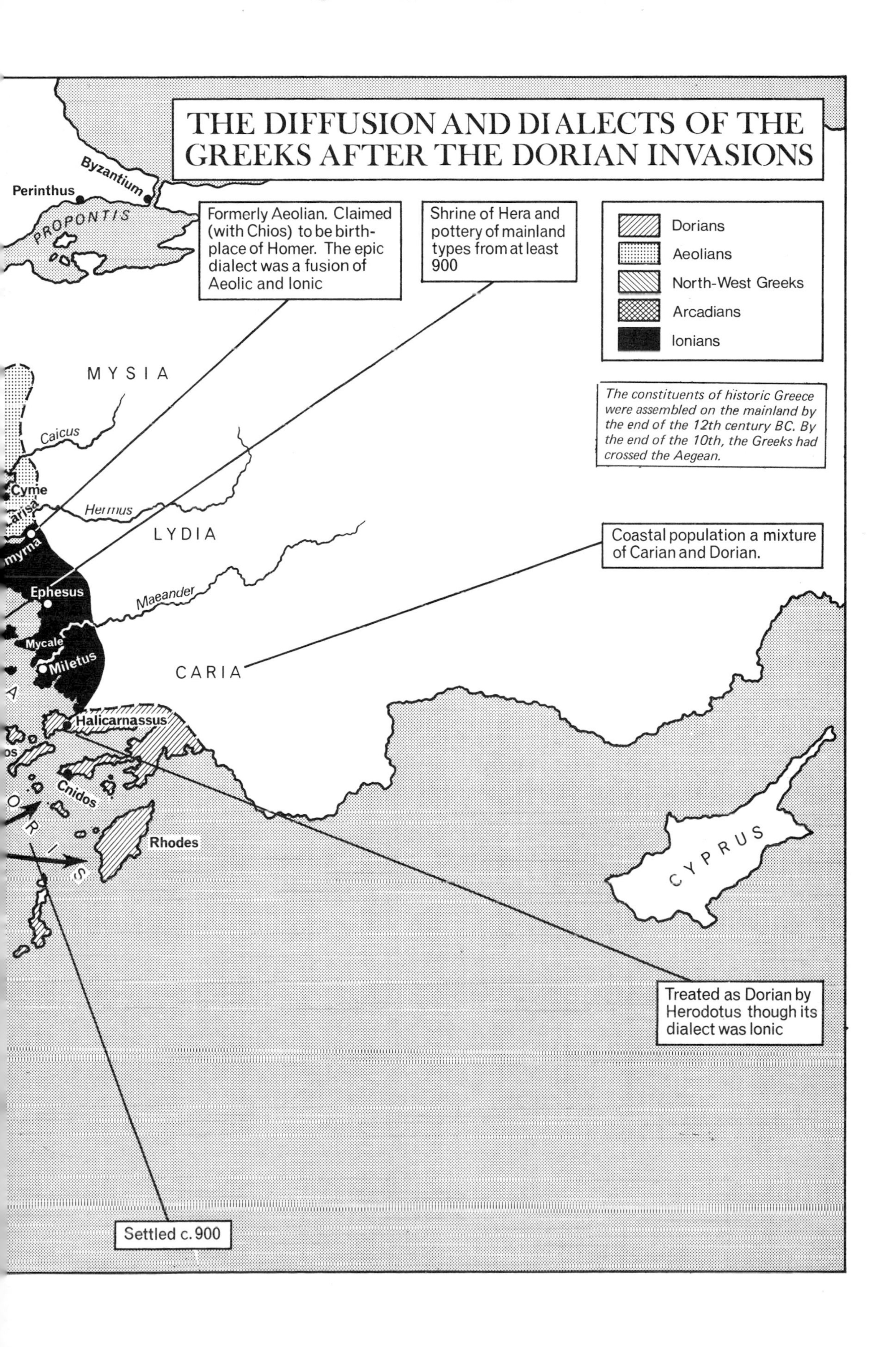
THE DIFFUSION AND DIALECTS OF THE GREEKS AFTER THE DORIAN INVASIONS
Dorians
Aeolians
North-West Greeks
Arcadians
Ionians
The constituents of historic Greece were assembled on the mainland by the end of the 12th century BC. By the end of the 10th, the Greeks had crossed the Aegean.
Formerly Aeolian. Claimed (with Chios) to be birth-place of Homer. The epic dialect was a fusion of Aeolic and Ionic
Shrine of Hera and pottery of mainland types from at least 900
Coastal population a mixture of Carian and Dorian.
Treated as Dorian by Herodotus though its dialect was Ionic
Settled c. 900
Byzantium
Perinthus
PROPONTIS
MYSIA
Caicus
Cyme
Hermus
LYDIA
Ephesus
Maeander
Mycale
Miletus
CARIA
Halicarnassus
Cnidos
Rhodes
CYPRUS

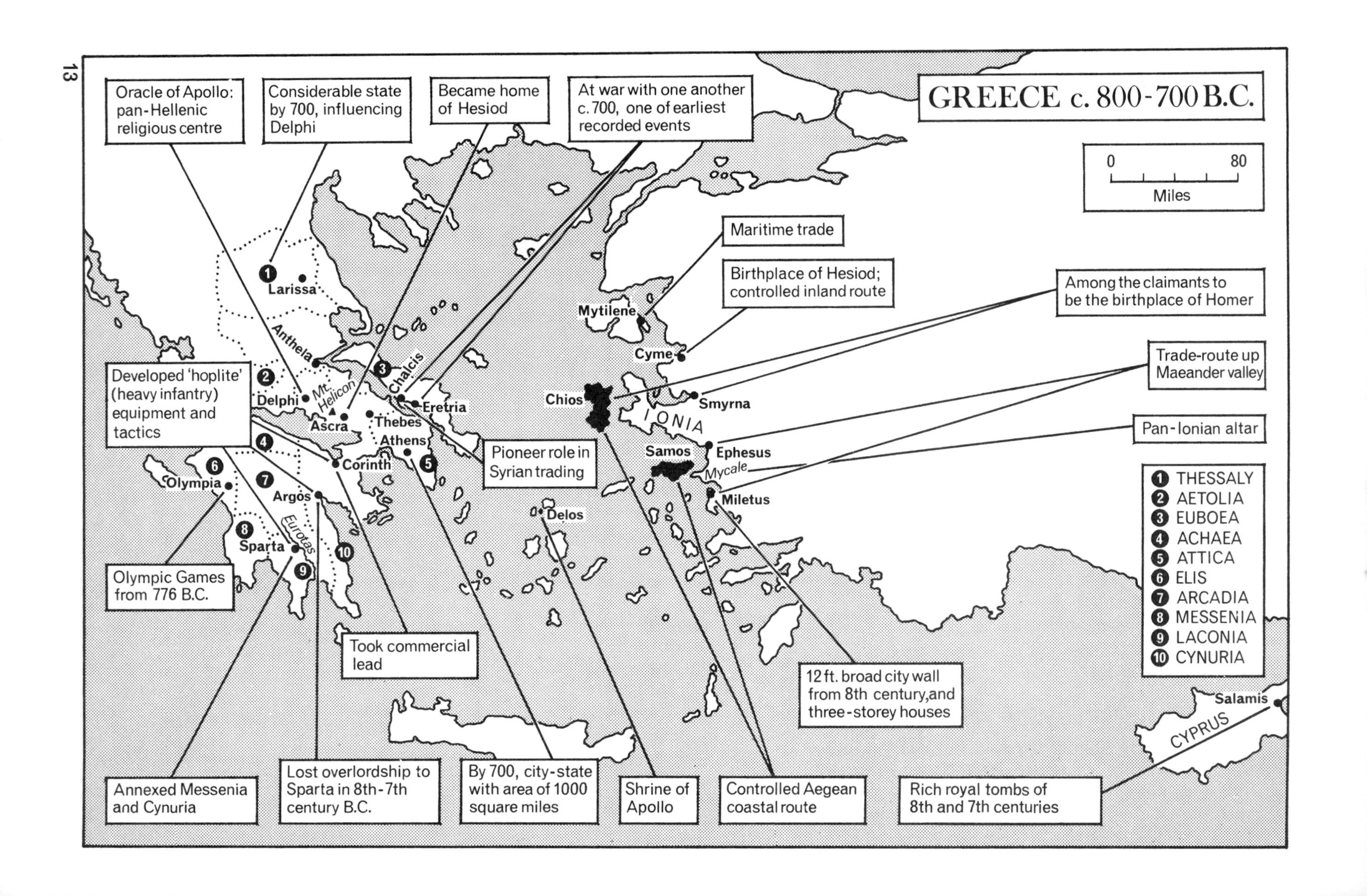
GREECE c. 800-700 B.C.
0 80
Miles
Oracle of Apollo: pan-Hellenic religious centre
Considerable state by 700, influencing Delphi
Became home of Hesiod
At war with one another c. 700, one of earliest recorded events
Maritime trade
Birthplace of Hesiod; controlled inland route
Among the claimants to be the birthplace of Homer
Trade-route up Maeander valley
Pan-Ionian altar
Developed 'hoplite' (heavy infantry) equipment and tactics
Pioneer role in Syrian trading
Olympic Games from 776 B.C.
Took commercial lead
12 ft. broad city wall from 8th century, and three-storey houses
Annexed Messenia and Cynuria
Lost overlordship to Sparta in 8th-7th century B.C.
By 700, city-state with area of 1000 square miles
Shrine of Apollo
Controlled Aegean coastal route
Rich royal tombs of 8th and 7th centuries
1 THESSALY
2 AETOLIA
3 EUBOEA
4 ACHAEA
5 ATTICA
6 ELIS
7 ARCADIA
8 MESSENIA
9 LACONIA
10 CYNURIA
Larissa
Anthela
Delphi
Mt. Helicon
Ascra
Thebes
Chalcis
Eretria
Athens
Corinth
Olympia
Argos
Sparta
Eurotas
Mytilene
Cyme
Chios
Smyrna
IONIA
Samos
Ephesus
Mycale
Miletus
Delos
Salamis
CYPRUS

THE PRINCIPAL AGRICULTURAL PRODUCTS OF GREECE

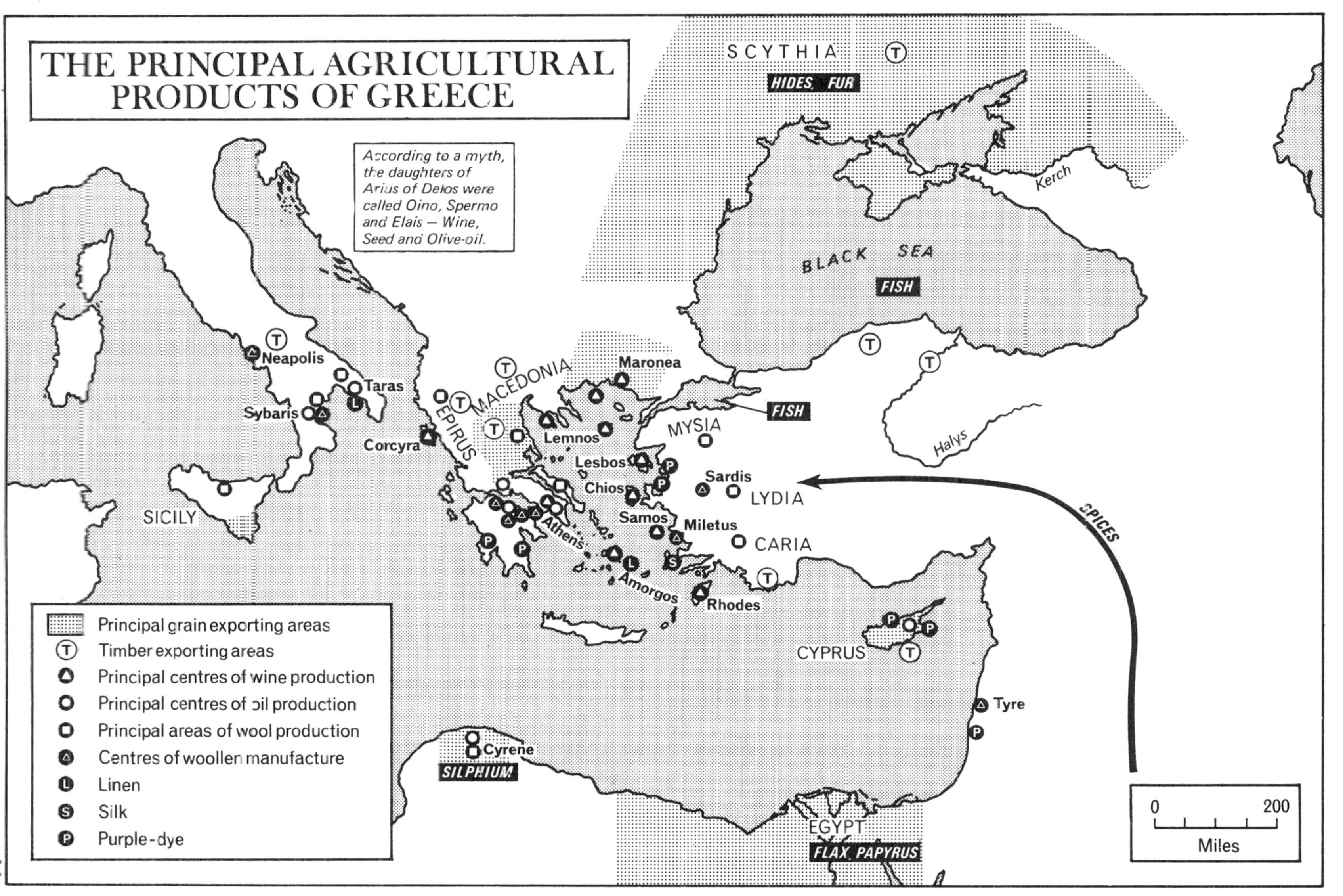

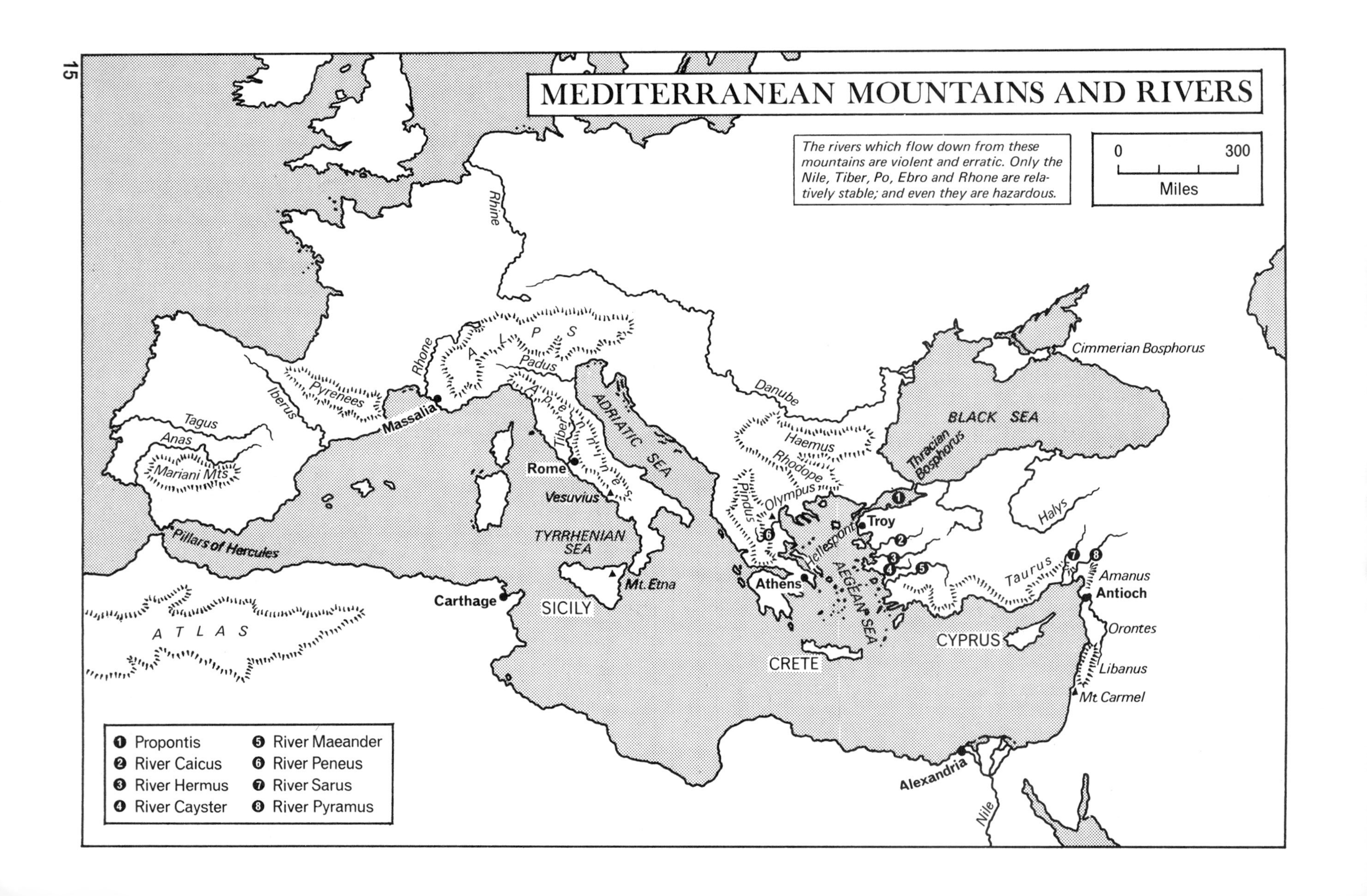
MEDITERRANEAN MOUNTAINS AND RIVERS
The rivers which flow down from these mountains are violent and erratic. Only the Nile, Tiber, Po, Ebro and Rhone are relatively stable; and even they are hazardous.
0
300
Miles
Rhine
Rhone
A L P S
Padus
Pyrenees
Iberus
Tagus
Anas
Mariani Mts
Massalia
Pillars of Hercules
A T L A S
Carthage
Rome
Tiber
Apennines
Vesuvius
TYRRHENIAN SEA
ADRIATIC SEA
SICILY
Mt Etna
Danube
Haemus
Rhodope
Pindus
Olympus
Hellespont
Troy
Athens
AEGEAN SEA
CRETE
BLACK SEA
Thracian Bosphorus
Cimmerian Bosphorus
Halys
Taurus
Amanus
Antioch
Orontes
CYPRUS
Libanus
Mt Carmel
Alexandria
Nile
1 Propontis
2 River Caicus
3 River Hermus
4 River Cayster
5 River Maeander
6 River Peneus
7 River Sarus
8 River Pyramus

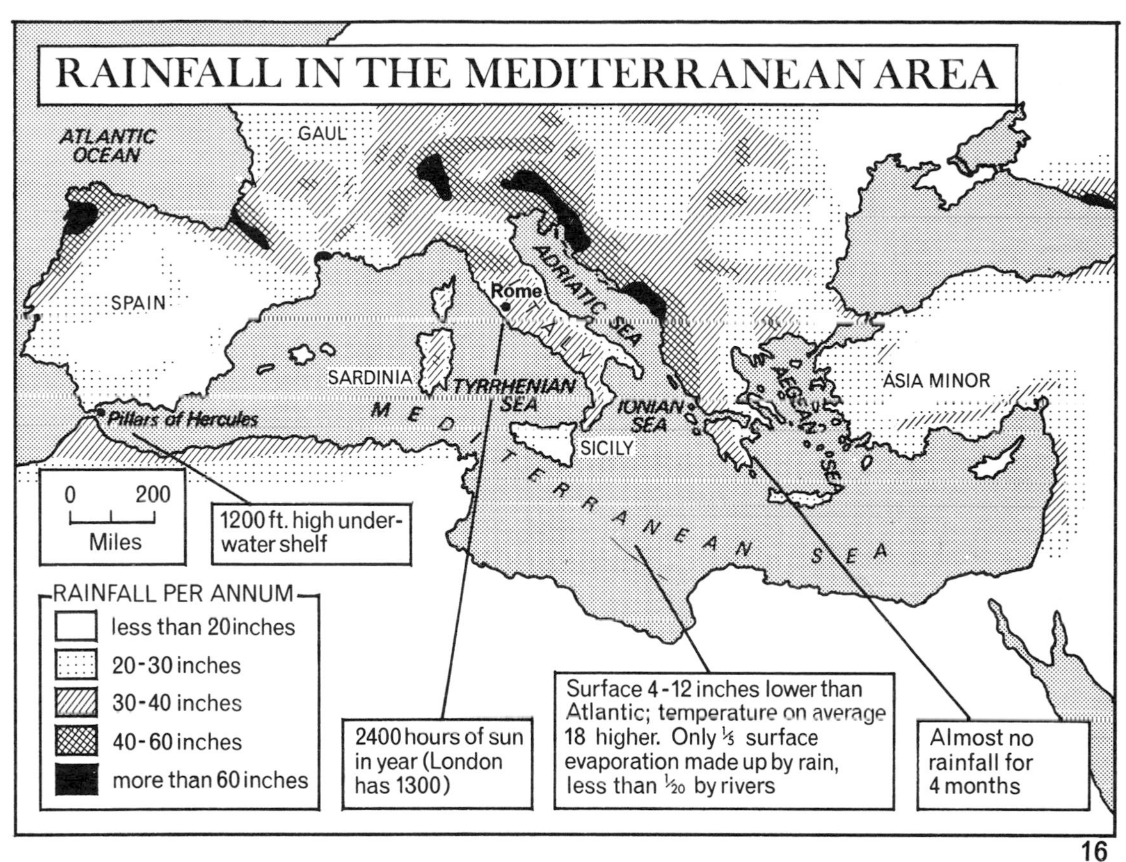
RAINFALL IN THE MEDITERRANEAN AREA
ATLANTIC OCEAN
GAUL
SPAIN
Rome
ADRIATIC SEA
ITALY
SARDINIA
TYRRHENIAN SEA
IONIAN SEA
SICILY
AEGEAN SEA
ASIA MINOR
Pillars of Hercules
MEDITERRANEAN SEA
0 200
Miles
1200 ft. high under-water shelf
RAINFALL PER ANNUM
less than 20 inches
20-30 inches
30-40 inches
40-60 inches
more than 60 inches
2400 hours of sun in year (London has 1300)
Surface 4-12 inches lower than Atlantic; temperature on average 18 higher. Only ⅕ surface evaporation made up by rain, less than 1/20 by rivers
Almost no rainfall for 4 months

16

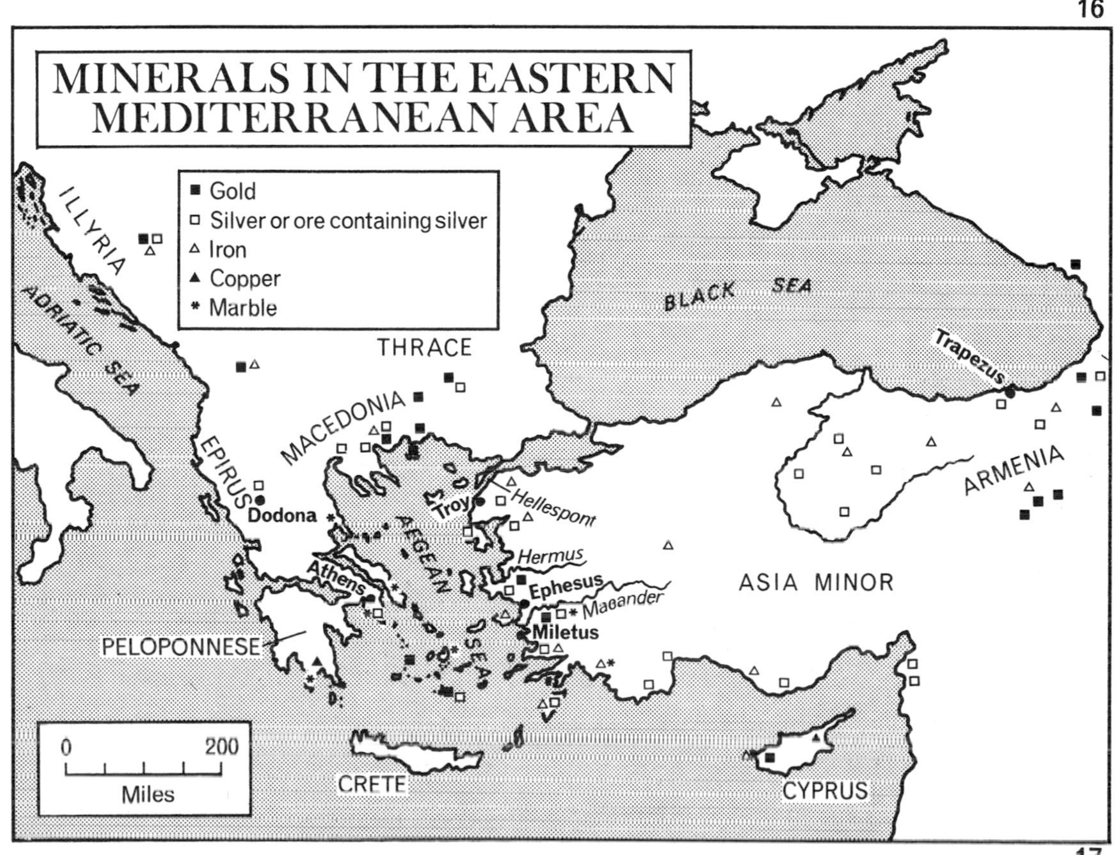
MINERALS IN THE EASTERN MEDITERRANEAN AREA
Gold
Silver or ore containing silver
Iron
Copper
Marble
ILLYRIA
ADRIATIC SEA
THRACE
BLACK SEA
Trapezus
MACEDONIA
EPIRUS
Dodona
Troy
Hellespont
AEGEAN SEA
Hermus
Athens
Ephesus
Maeander
Miletus
ARMENIA
ASIA MINOR
PELOPONNESE
0 200
Miles
CRETE
CYPRUS

17

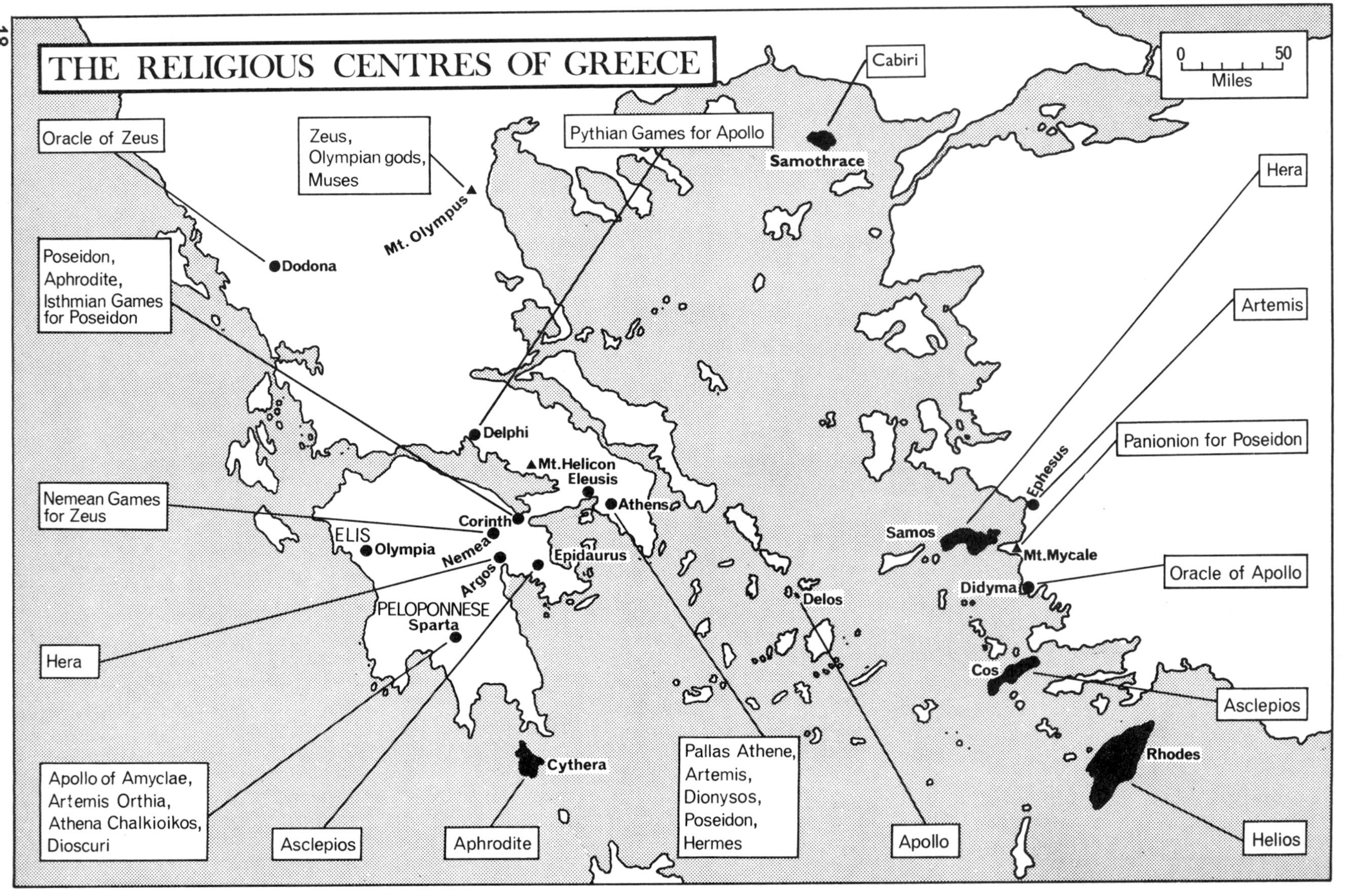
THE RELIGIOUS CENTRES OF GREECE
0
50
Miles
Oracle of Zeus
Dodona
Zeus,
Olympian gods,
Muses
Mt. Olympus
Pythian Games for Apollo
Cabiri
Samothrace
Poseidon,
Aphrodite,
Isthmian Games
for Poseidon
Delphi
Mt.Helicon
Eleusis
Athens
Nemean Games
for Zeus
Corinth
ELIS
Olympia
Nemea
Argos
Epidaurus
PELOPONNESE
Sparta
Hera
Apollo of Amyclae,
Artemis Orthia,
Athena Chalkioikos,
Dioscuri
Asclepios
Aphrodite
Cythera
Pallas Athene,
Artemis,
Dionysos,
Poseidon,
Hermes
Delos
Apollo
Hera
Artemis
Panionion for Poseidon
Ephesus
Samos
Mt.Mycale
Oracle of Apollo
Didyma
Cos
Asclepios
Rhodes
Helios

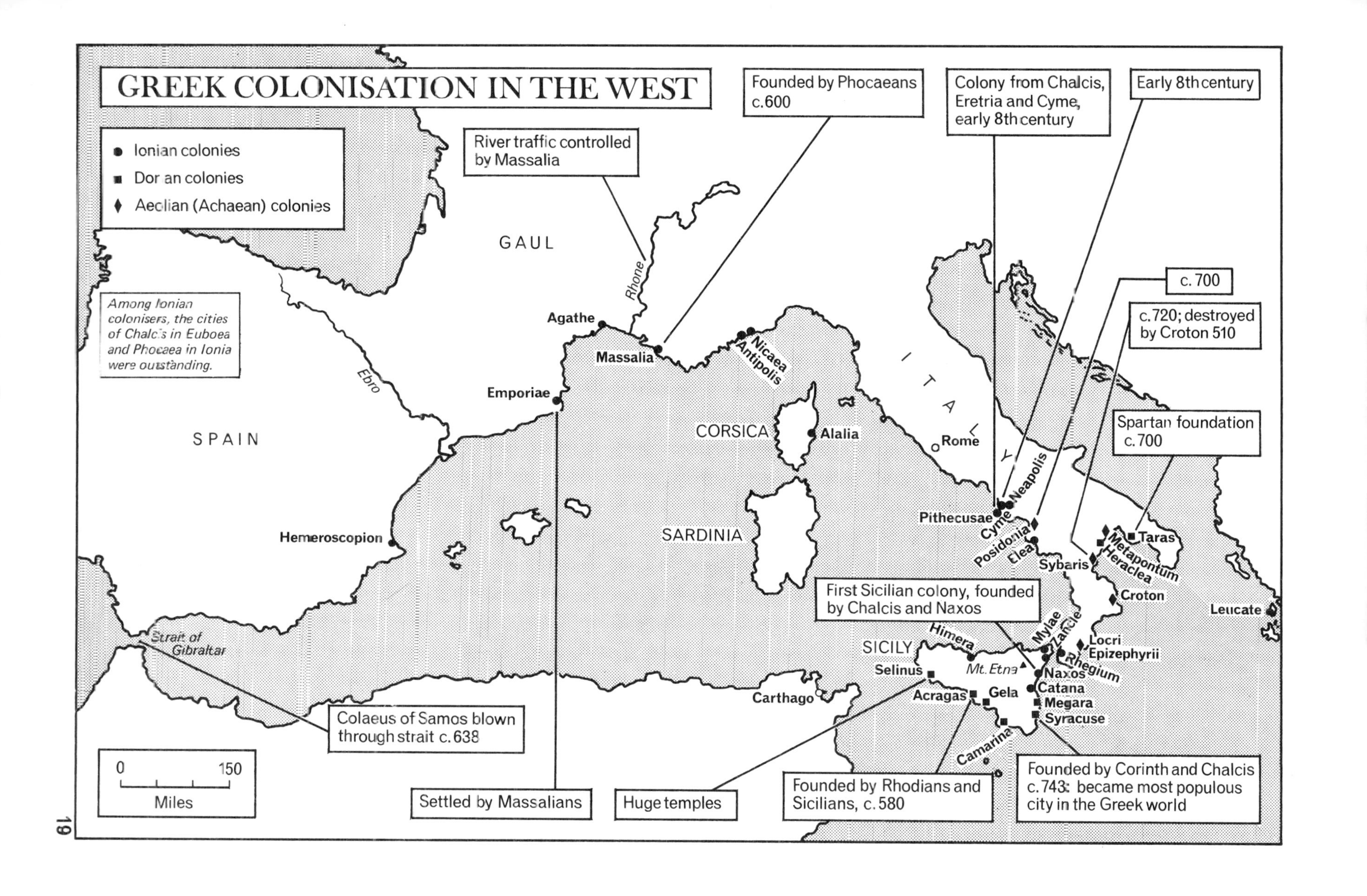
GREEK COLONISATION IN THE WEST
Ionian colonies
Dorian colonies
Aeolian (Achaean) colonies
Among Ionian colonisers, the cities of Chalcis in Euboea and Phocaea in Ionia were outstanding.
River traffic controlled by Massalia
Founded by Phocaeans c.600
Colony from Chalcis, Eretria and Cyme, early 8th century
Early 8th century
c.700
c.720; destroyed by Croton 510
Spartan foundation c.700
First Sicilian colony, founded by Chalcis and Naxos
Colaeus of Samos blown through strait c.638
Settled by Massalians
Huge temples
Founded by Rhodians and Sicilians, c.580
Founded by Corinth and Chalcis c.743: became most populous city in the Greek world
0
150
Miles
GAUL
SPAIN
CORSICA
SARDINIA
SICILY
ITALY
Rhone
Ebro
Strait of Gibraltar
Agathe
Massalia
Emporiae
Hemeroscopion
Nicaea
Antipolis
Alalia
Rome
Carthago
Pithecusae
Cyme
Neapolis
Posidonia
Elea
Sybaris
Taras
Metapontum
Heraclea
Croton
Leucate
Locri
Epizephyrii
Rhegium
Mylae
Zancle
Naxos
Catana
Megara
Syracuse
Himera
Mt. Etna
Selinus
Acragas
Gela
Camarina

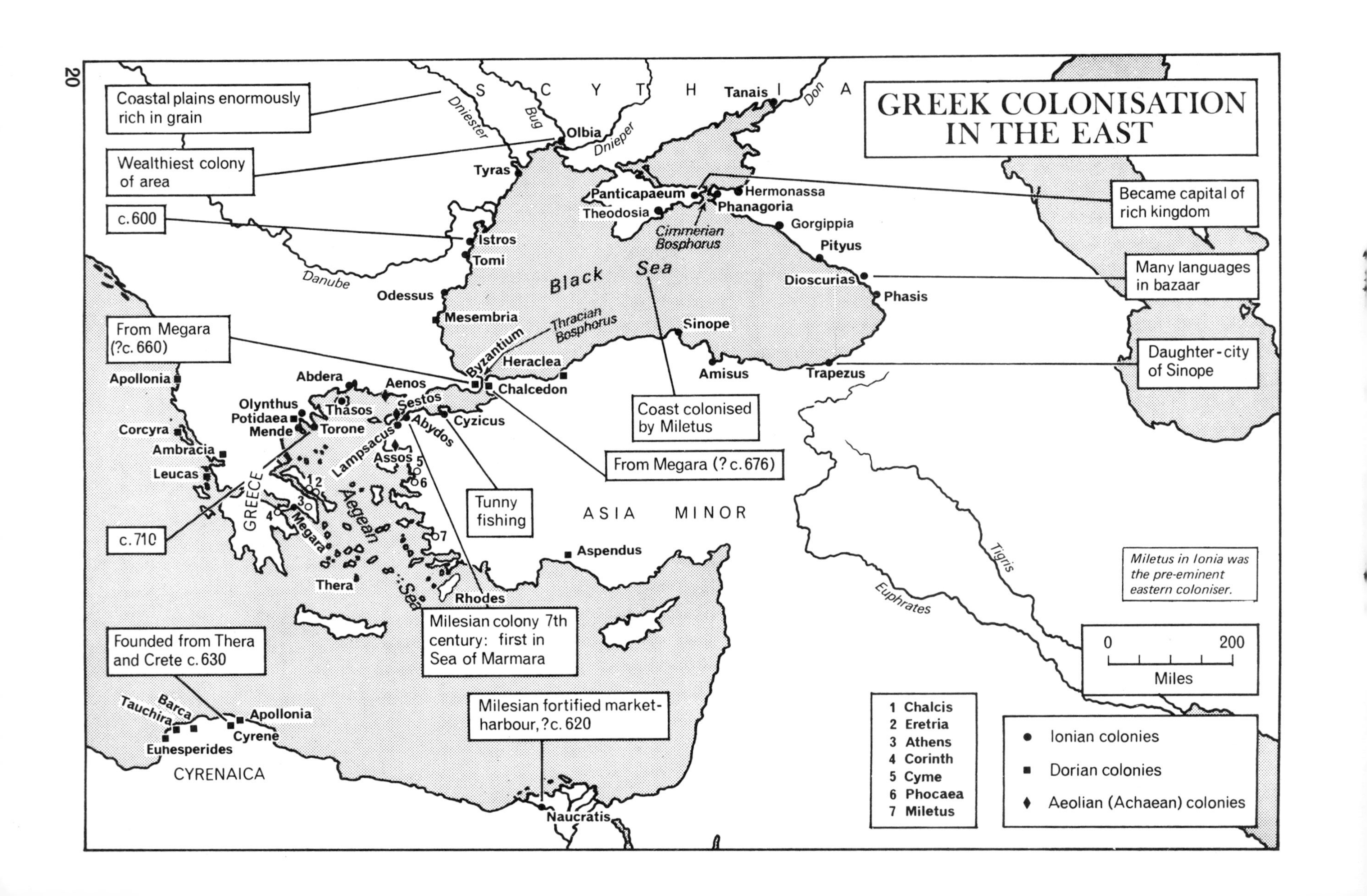
GREEK COLONISATION IN THE EAST
Coastal plains enormously rich in grain
Wealthiest colony of area
c.600
From Megara (?c.660)
c.710
Founded from Thera and Crete c.630
Milesian colony 7th century: first in Sea of Marmara
Milesian fortified market-harbour, ?c.620
Tunny fishing
From Megara (?c.676)
Coast colonised by Miletus
Became capital of rich kingdom
Many languages in bazaar
Daughter-city of Sinope
Miletus in Ionia was the pre-eminent eastern coloniser.
0
200
Miles
1 Chalcis
2 Eretria
3 Athens
4 Corinth
5 Cyme
6 Phocaea
7 Miletus
Ionian colonies
Dorian colonies
Aeolian (Achaean) colonies
S C Y T H I A
ASIA MINOR
GREECE
CYRENAICA
Black Sea
Aegean Sea
Thracian Bosphorus
Cimmerian Bosphorus
Dniester
Bug
Dnieper
Don
Danube
Tigris
Euphrates
Tanais
Olbia
Tyras
Panticapaeum
Hermonassa
Theodosia
Phanagoria
Gorgippia
Pityus
Dioscurias
Phasis
Istros
Tomi
Odessus
Mesembria
Byzantium
Heraclea
Sinope
Amisus
Trapezus
Chalcedon
Cyzicus
Abdera
Aenos
Sestos
Abydos
Lampsacus
Assos
Thasos
Olynthus
Potidaea
Mende
Torone
Megara
Thera
Rhodes
Aspendus
Apollonia
Corcyra
Ambracia
Leucas
Barca
Tauchira
Euhesperides
Cyrene
Naucratis

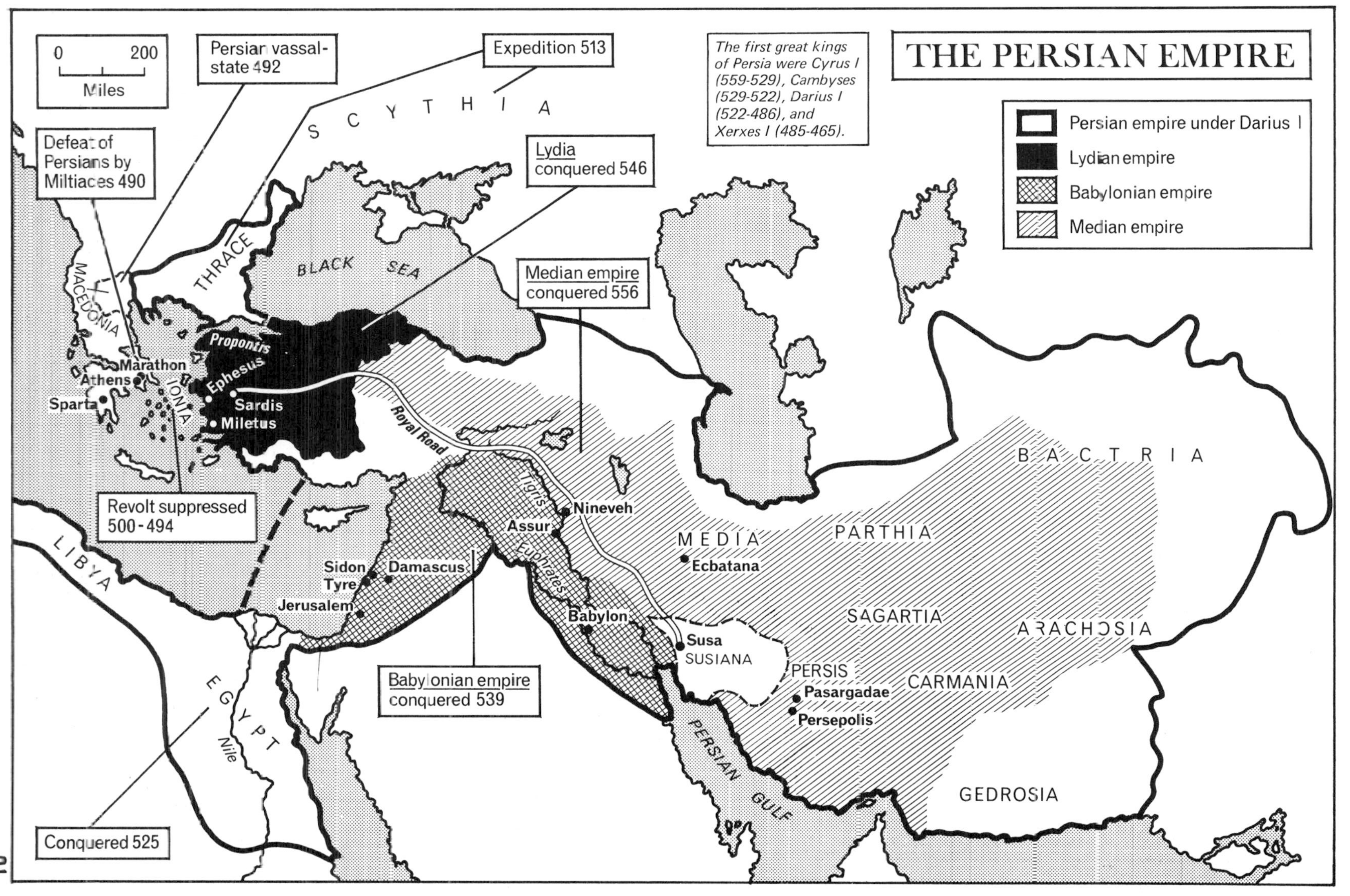
THE PERSIAN EMPIRE
Persian empire under Darius I
Lydian empire
Babylonian empire
Median empire
The first great kings of Persia were Cyrus I (559-529), Cambyses (529-522), Darius I (522-486), and Xerxes I (485-465).
0 200
Miles
Persian vassal-state 492
Expedition 513
Defeat of Persians by Miltiades 490
Lydia conquered 546
Median empire conquered 556
Revolt suppressed 500-494
Babylonian empire conquered 539
Conquered 525
SCYTHIA
THRACE
BLACK SEA
MACEDONIA
Proponitis
Marathon
Athens
Sparta
IONIA
Ephesus
Sardis
Miletus
Royal Road
Tigris
Euphrates
Nineveh
Assur
Babylon
Susa
SUSIANA
MEDIA
Ecbatana
PARTHIA
SAGARTIA
ARACHOSIA
BACTRIA
PERSIS
Pasargadae
Persepolis
CARMANIA
GEDROSIA
PERSIAN GULF
Sidon
Tyre
Damascus
Jerusalem
LIBYA
EGYPT
Nile

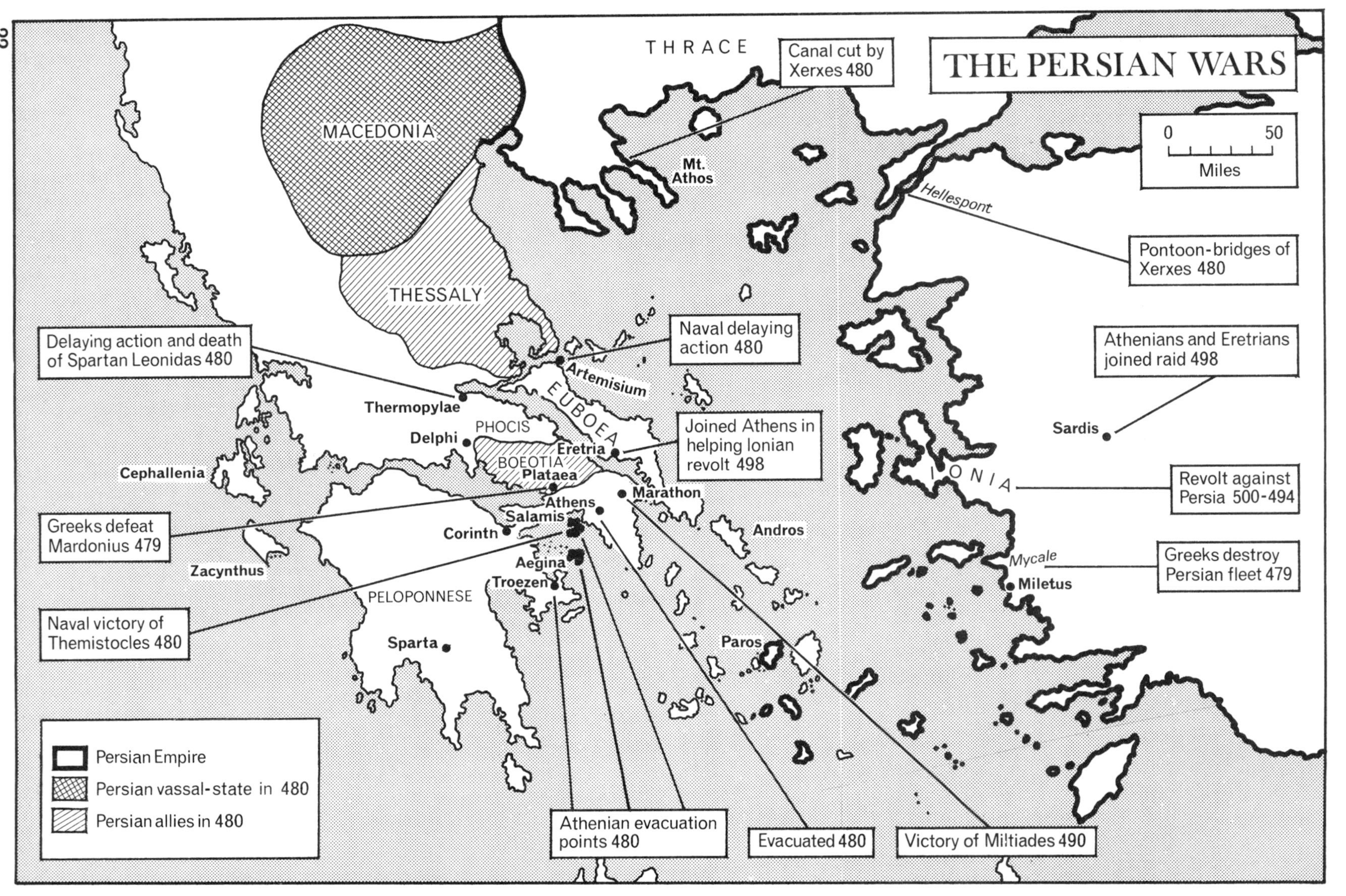
THE PERSIAN WARS
0 50
Miles
THRACE
Canal cut by Xerxes 480
Mt. Athos
Hellespont
Pontoon-bridges of Xerxes 480
MACEDONIA
THESSALY
Naval delaying action 480
Artemisium
EUBOEA
Delaying action and death of Spartan Leonidas 480
Thermopylae
PHOCIS
Delphi
BOEOTIA
Eretria
Plataea
Joined Athens in helping Ionian revolt 498
Athenians and Eretrians joined raid 498
Sardis
IONIA
Revolt against Persia 500-494
Cephallenia
Marathon
Athens
Salamis
Corinth
Andros
Greeks defeat Mardonius 479
Zacynthus
Aegina
Troezen
Mycale
Miletus
Greeks destroy Persian fleet 479
PELOPONNESE
Naval victory of Themistocles 480
Sparta
Paros
Persian Empire
Persian vassal-state in 480
Persian allies in 480
Athenian evacuation points 480
Evacuated 480
Victory of Miltiades 490

THE BATTLE OF SALAMIS 480 B.C.

N.B. There are also other versions of this battle

The naval victory of the Greeks over the Persians

'But those Greek ships,
Skilfully handled, kept the outer station
Ringing us round and striking in, till ships
Turned turtle, and you could not see the water
For blood and wreckage . . .'
(Aeschylus, The Persians, 417-20, Translated by A.R. Burn)

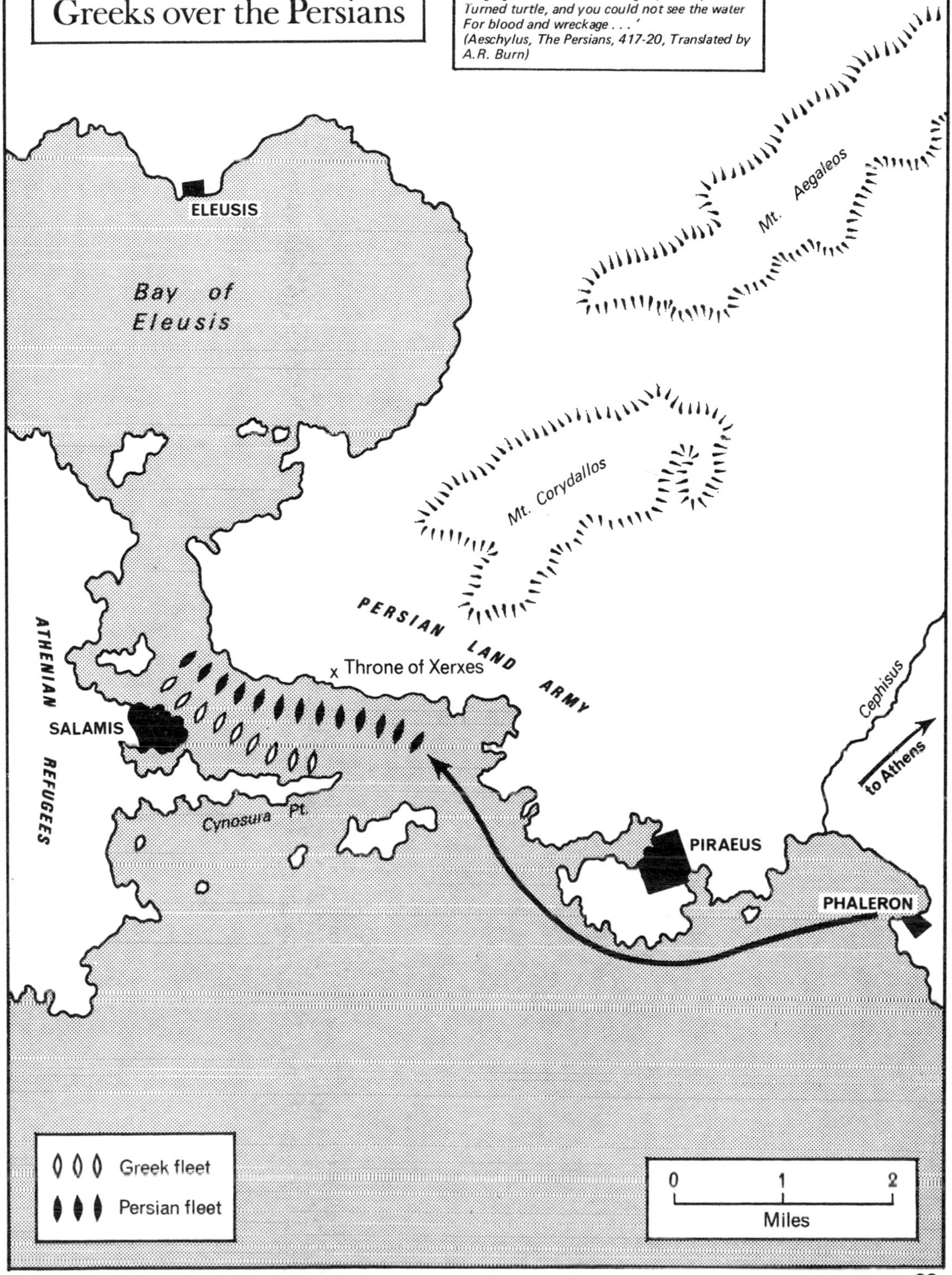

ELEUSIS
Bay of Eleusis
Mt. Aegaleos
Mt. Corydallos
PERSIAN LAND ARMY
x Throne of Xerxes
ATHENIAN REFUGEES
SALAMIS
Cynosura Pt.
Cephisus
to Athens
PIRAEUS
PHALERON

Greek fleet
Persian fleet

0 1 2
Miles

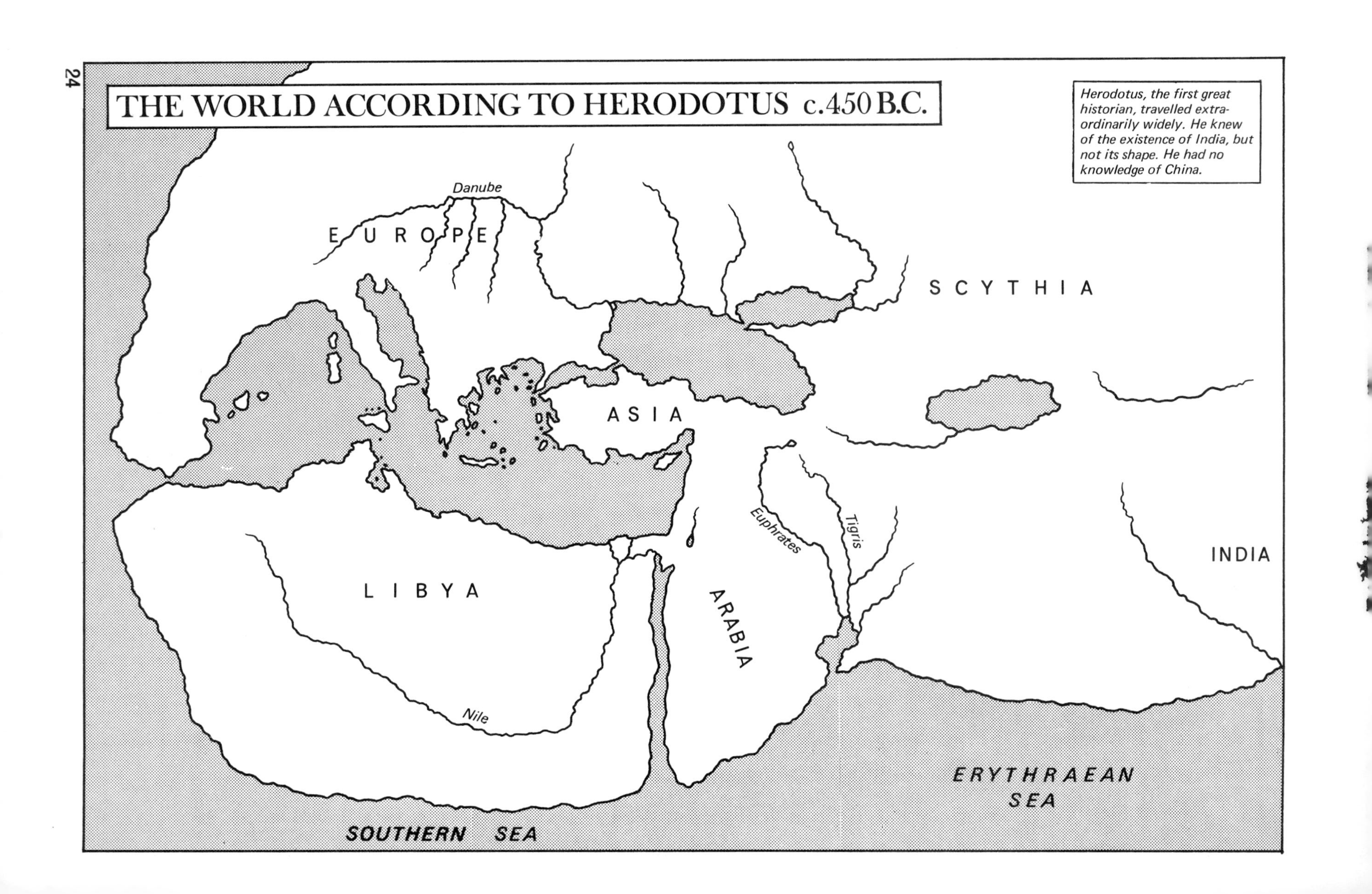
THE WORLD ACCORDING TO HERODOTUS c.450 B.C.
Herodotus, the first great historian, travelled extraordinarily widely. He knew of the existence of India, but not its shape. He had no knowledge of China.
Danube
EUROPE
SCYTHIA
ASIA
Euphrates
Tigris
INDIA
LIBYA
ARABIA
Nile
ERYTHRAEAN SEA
SOUTHERN SEA

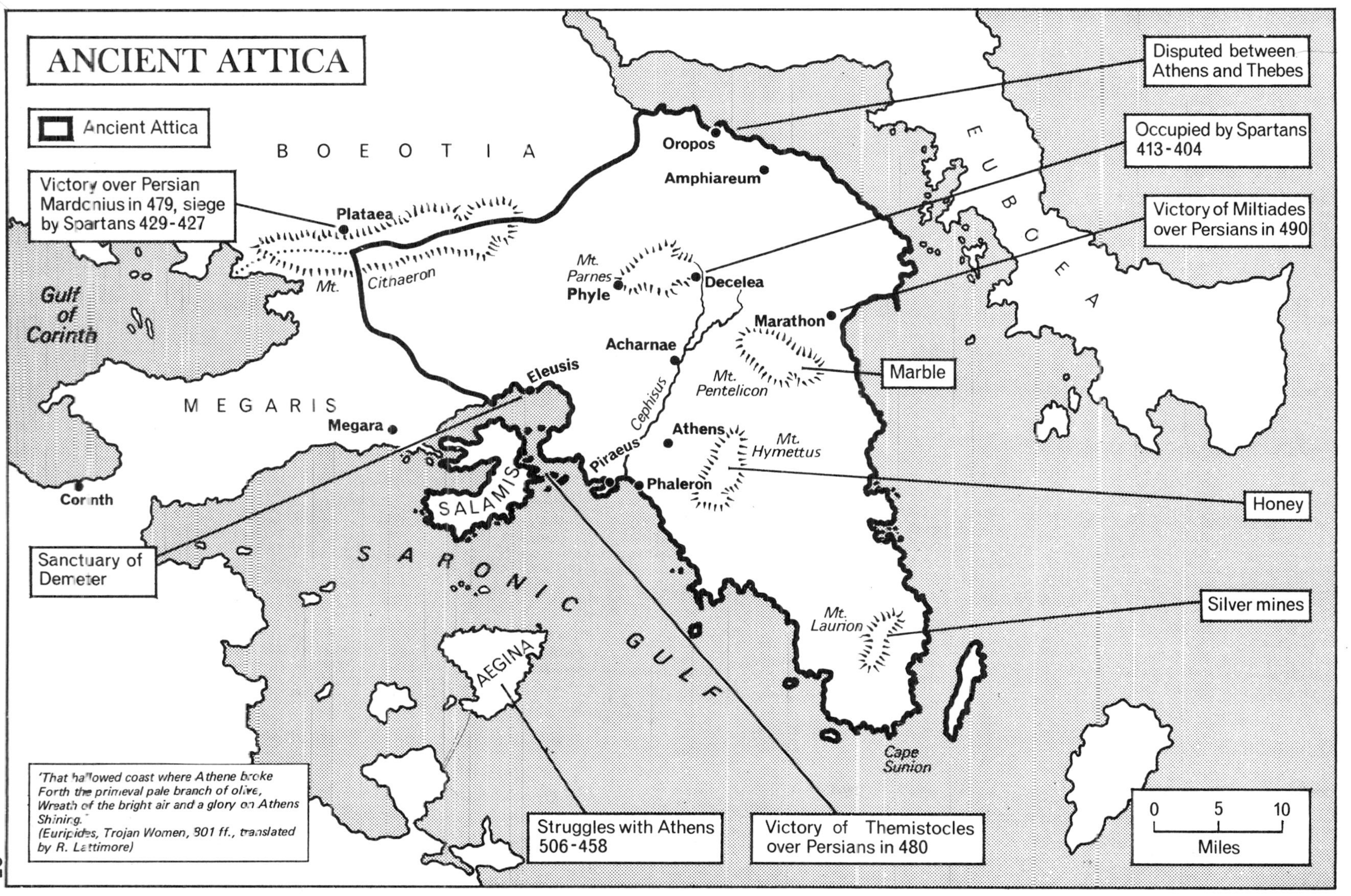

ANCIENT ATTICA
Ancient Attica
BOEOTIA
Victory over Persian Mardonius in 479, siege by Spartans 429-427
Plataea
Mt. Cithaeron
Gulf of Corinth
Corinth
MEGARIS
Megara
Eleusis
Sanctuary of Demeter
SALAMIS
SARONIC GULF
AEGINA
Struggles with Athens 506-458
Victory of Themistocles over Persians in 480
Oropos
Amphiareum
Disputed between Athens and Thebes
Mt. Parnes
Phyle
Decelea
Occupied by Spartans 413-404
Acharnae
Cephisus
Piraeus
Phaleron
Athens
Marathon
Victory of Miltiades over Persians in 490
Mt. Pentelicon
Marble
Mt. Hymettus
Honey
Mt. Laurion
Silver mines
Cape Sunion
EUBOEA
0 5 10
Miles
'That hallowed coast where Athene broke
Forth the primeval pale branch of olive,
Wreath of the bright air and a glory on Athens
Shining.'
(Euripides, Trojan Women, 801 ff., translated by R. Lattimore)

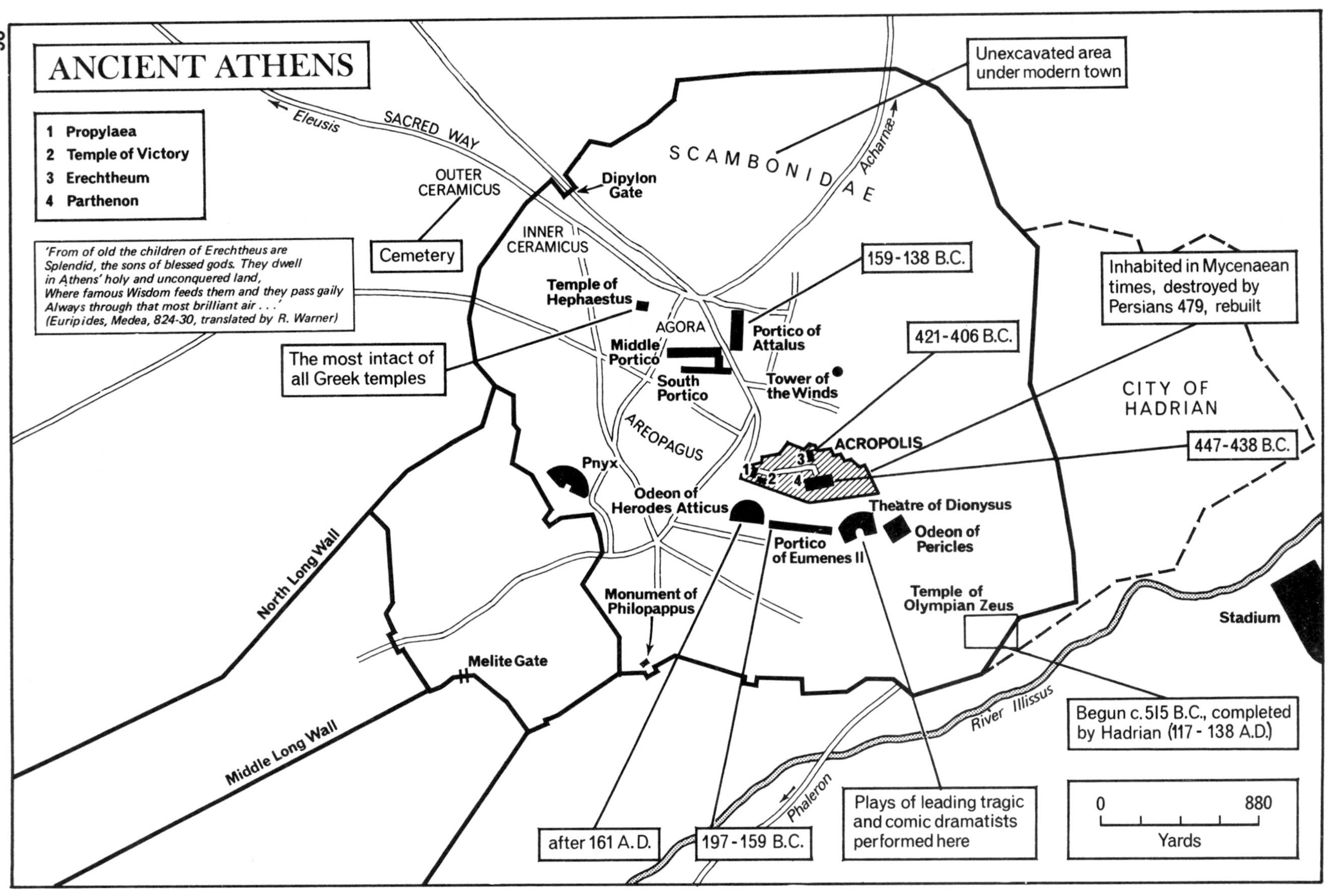

ANCIENT ATHENS
1 Propylaea
2 Temple of Victory
3 Erechtheum
4 Parthenon
'From of old the children of Erechtheus are Splendid, the sons of blessed gods. They dwell in Athens' holy and unconquered land, Where famous Wisdom feeds them and they pass gaily Always through that most brilliant air . . .'
(Euripides, Medea, 824-30, translated by R. Warner)
Eleusis
SACRED WAY
OUTER CERAMICUS
Cemetery
Dipylon Gate
SCAMBONIDAE
Acharnae
Unexcavated area under modern town
INNER CERAMICUS
Temple of Hephaestus
The most intact of all Greek temples
AGORA
Middle Portico
South Portico
Portico of Attalus
159-138 B.C.
Tower of the Winds
421-406 B.C.
Inhabited in Mycenaean times, destroyed by Persians 479, rebuilt
CITY OF HADRIAN
447-438 B.C.
ACROPOLIS
AREOPAGUS
Pnyx
Odeon of Herodes Atticus
Theatre of Dionysus
Odeon of Pericles
Portico of Eumenes II
Temple of Olympian Zeus
Stadium
Monument of Philopappus
Melite Gate
North Long Wall
Middle Long Wall
River Illissus
Begun c. 515 B.C., completed by Hadrian (117 - 138 A.D.)
Phaleron
after 161 A.D.
197 - 159 B.C.
Plays of leading tragic and comic dramatists performed here
0
880
Yards

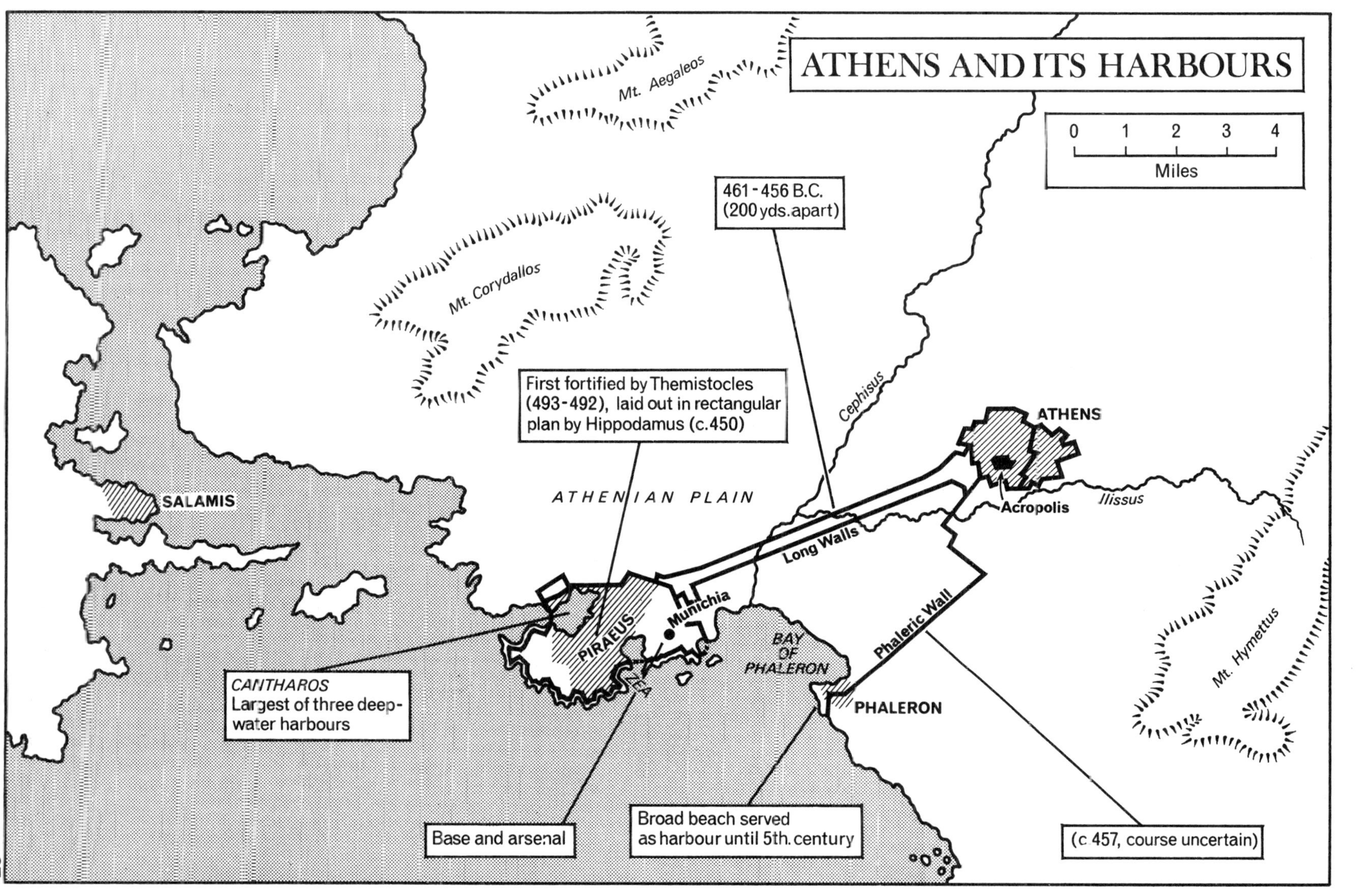
ATHENS AND ITS HARBOURS
0 1 2 3 4
Miles
Mt. Aegaleos
Mt. Corydallos
461 - 456 B.C.
(200 yds. apart)
First fortified by Themistocles
(493-492), laid out in rectangular
plan by Hippodamus (c.450)
Cephisus
ATHENS
Acropolis
Ilissus
ATHENIAN PLAIN
SALAMIS
Long Walls
Munichia
PIRAEUS
ZEA
BAY OF PHALERON
Phaleric Wall
PHALERON
Mt. Hymettus
CANTHAROS
Largest of three deep-
water harbours
Base and arsenal
Broad beach served
as harbour until 5th. century
(c 457, course uncertain)

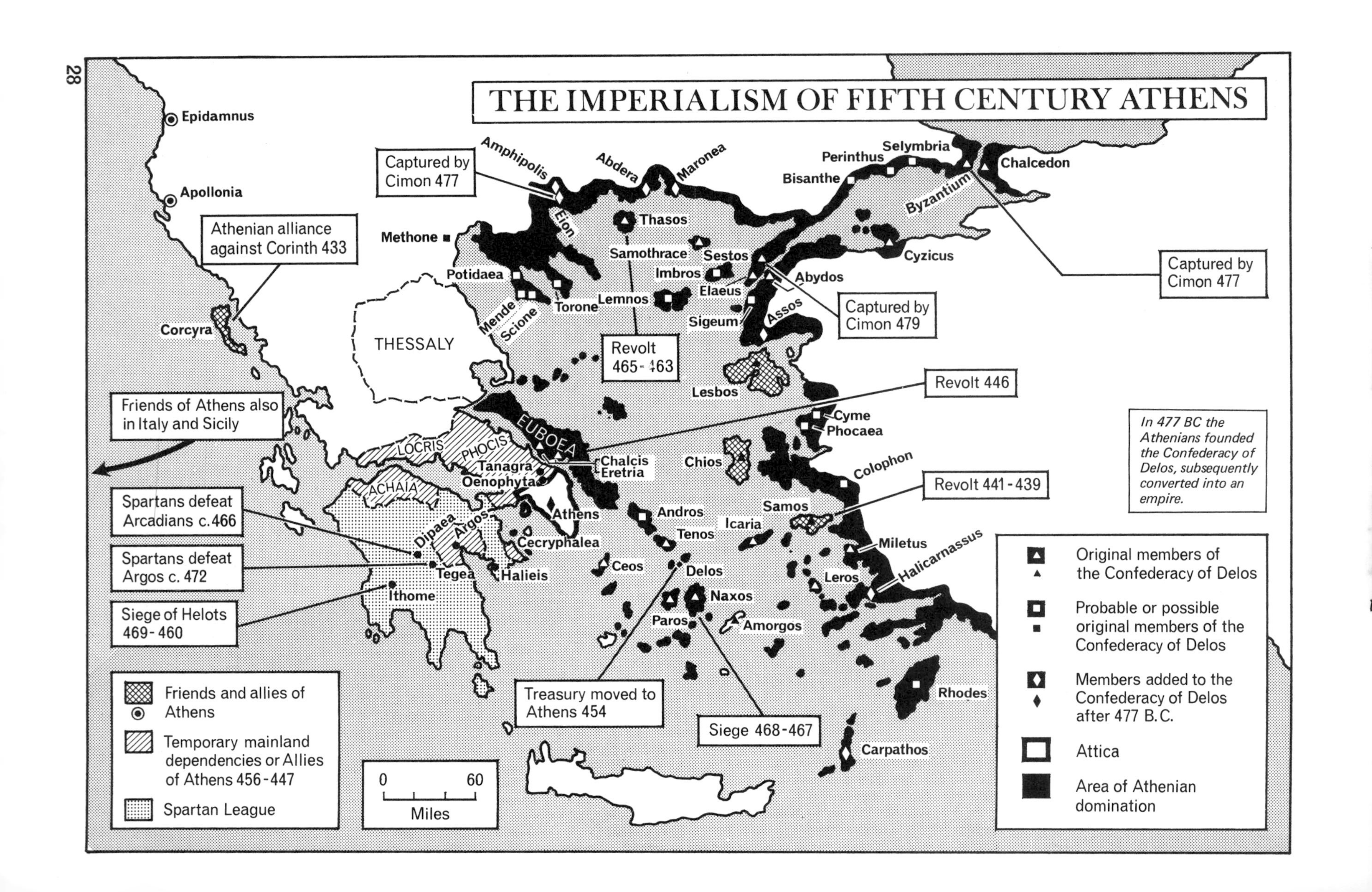

THE IMPERIALISM OF FIFTH CENTURY ATHENS
Epidamnus
Apollonia
Athenian alliance against Corinth 433
Corcyra
Friends of Athens also in Italy and Sicily
Captured by Cimon 477
Amphipolis
Eion
Abdera
Maronea
Methone
Potidaea
Mende
Scione
Torone
THESSALY
Thasos
Samothrace
Sestos
Imbros
Elaeus
Lemnos
Revolt 465-463
Sigeum
Assos
Abydos
Captured by Cimon 479
Lesbos
Bisanthe
Perinthus
Selymbria
Byzantium
Chalcedon
Cyzicus
Captured by Cimon 477
Revolt 446
Cyme
Phocaea
LOCRIS
PHOCIS
EUBOEA
Chalcis
Eretria
Tanagra
Oenophyta
ACHAIA
Chios
Colophon
Revolt 441-439
In 477 BC the Athenians founded the Confederacy of Delos, subsequently converted into an empire.
Spartans defeat Arcadians c. 466
Dipaea
Argos
Athens
Andros
Icaria
Samos
Tenos
Cecryphalea
Spartans defeat Argos c. 472
Tegea
Halieis
Ceos
Delos
Miletus
Halicarnassus
Leros
Ithome
Siege of Helots 469-460
Naxos
Paros
Amorgos
Treasury moved to Athens 454
Siege 468-467
Rhodes
Carpathos
Friends and allies of Athens
Temporary mainland dependencies or Allies of Athens 456-447
Spartan League
0
60
Miles
Original members of the Confederacy of Delos
Probable or possible original members of the Confederacy of Delos
Members added to the Confederacy of Delos after 477 B.C.
Attica
Area of Athenian domination

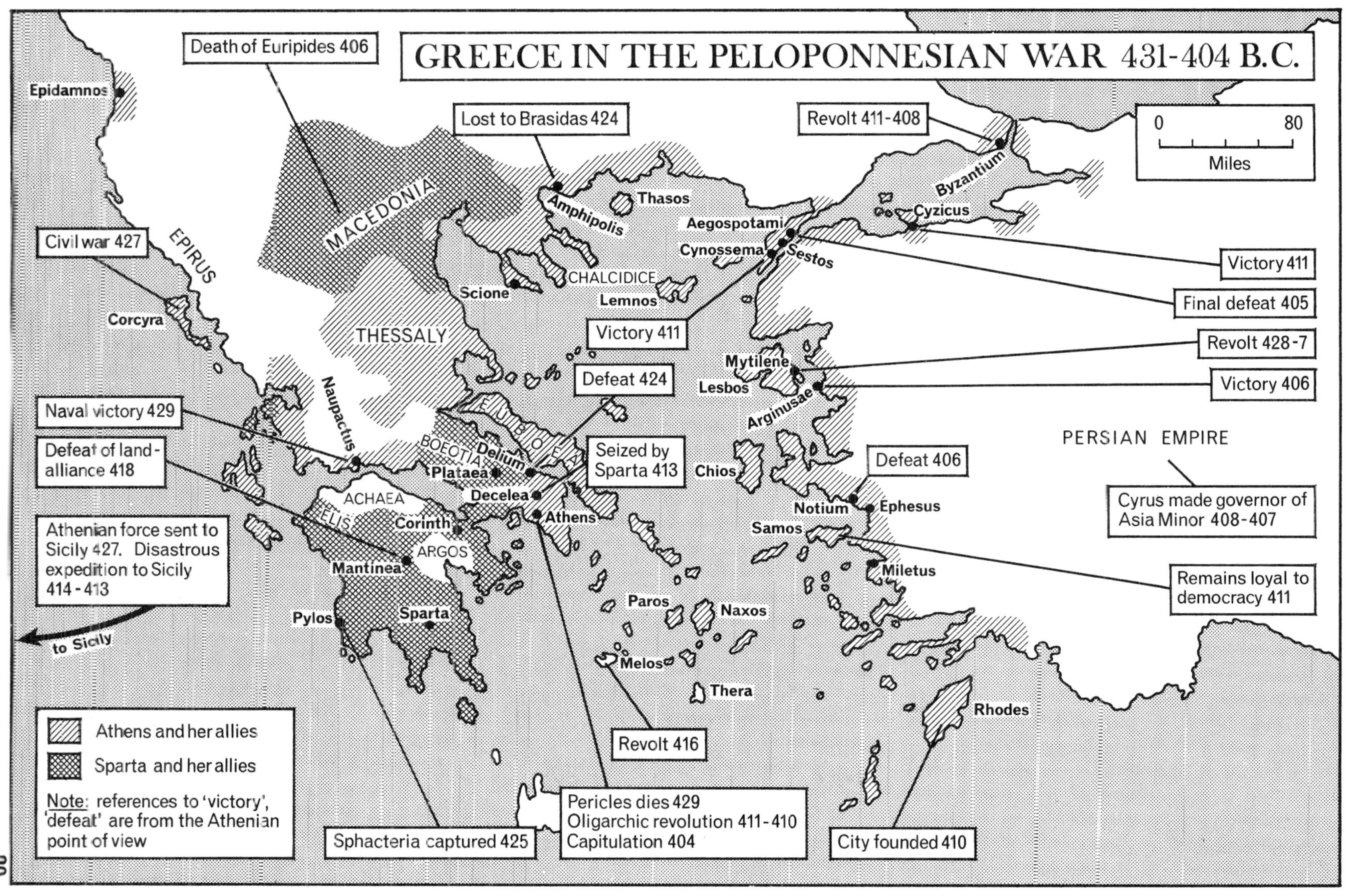

GREECE IN THE PELOPONNESIAN WAR 431-404 B.C.
Death of Euripides 406
Epidamnos
Lost to Brasidas 424
Revolt 411-408
0
80
Miles
Byzantium
Cyzicus
Thasos
MACEDONIA
Amphipolis
Aegospotami
Cynossema
Sestos
Civil war 427
EPIRUS
Victory 411
Scione
CHALCIDICE
Lemnos
Final defeat 405
Corcyra
THESSALY
Victory 411
Revolt 428-7
Mytilene
Defeat 424
Lesbos
Victory 406
Naupactus
Naval victory 429
Arginusae
EUBOEA
BOEOTIA
Delium
Seized by Sparta 413
PERSIAN EMPIRE
Defeat of land-alliance 418
Plataea
Chios
Defeat 406
ACHAEA
Decelea
Notium
Ephesus
ELIS
Corinth
Athens
Cyrus made governor of Asia Minor 408-407
Athenian force sent to Sicily 427. Disastrous expedition to Sicily 414-413
ARGOS
Samos
Mantinea
Miletus
Remains loyal to democracy 411
Paros
Naxos
Pylos
Sparta
to Sicily
Melos
Thera
Rhodes
Athens and her allies
Sparta and her allies
Revolt 416
Note: references to 'victory', 'defeat' are from the Athenian point of view
Sphacteria captured 425
Pericles dies 429
Oligarchic revolution 411-410
Capitulation 404
City founded 410

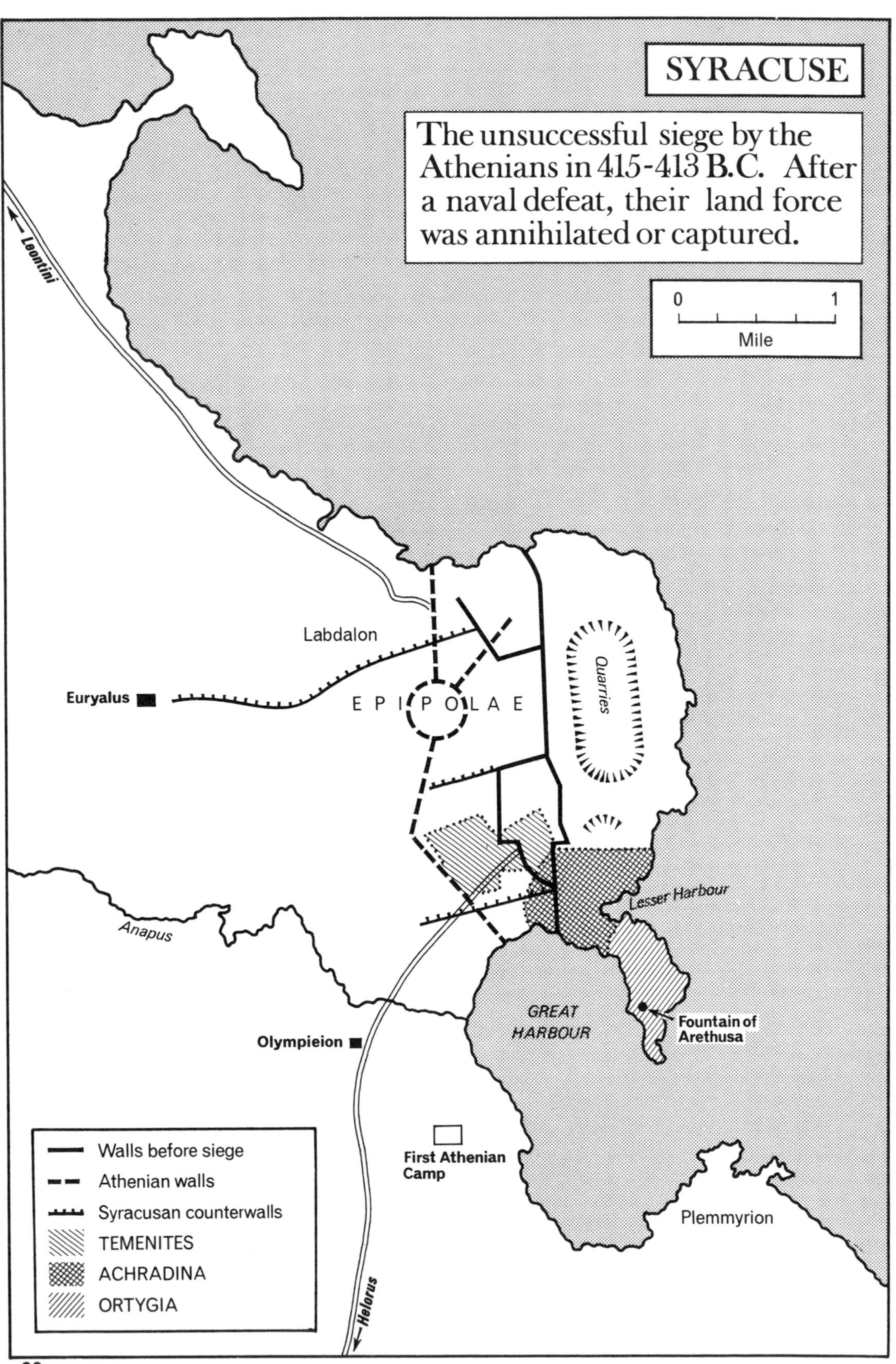
SYRACUSE
The unsuccessful siege by the Athenians in 415-413 B.C. After a naval defeat, their land force was annihilated or captured.
0
1
Mile
Leontini
Labdalon
Euryalus
E P I P O L A E
Quarries
Lesser Harbour
Anapus
GREAT HARBOUR
Fountain of Arethusa
Olympieion
First Athenian Camp
Plemmyrion
Helorus
Walls before siege
Athenian walls
Syracusan counterwalls
TEMENITES
ACHRADINA
ORTYGIA

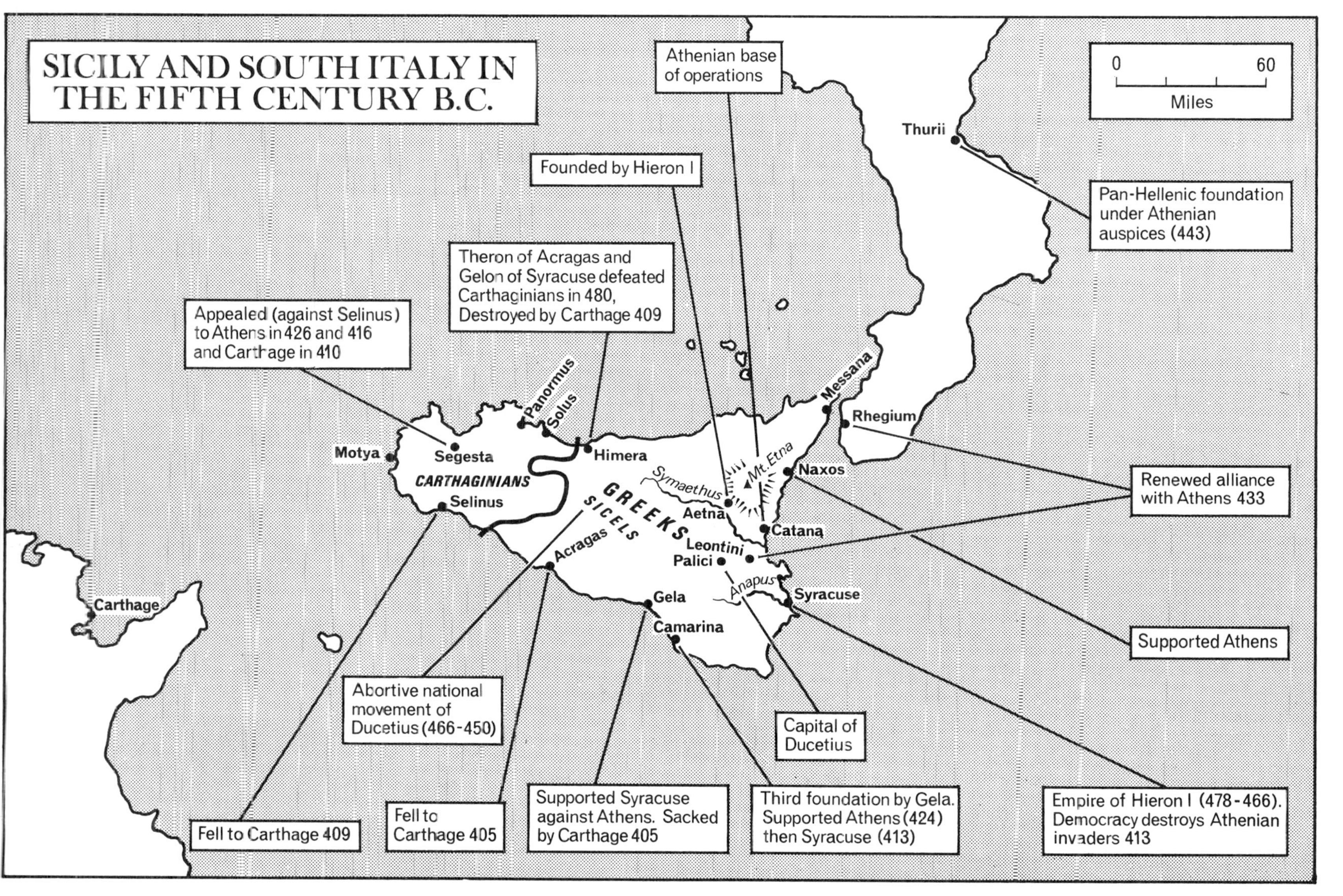
SICILY AND SOUTH ITALY IN THE FIFTH CENTURY B.C.
0
60
Miles
Athenian base of operations
Founded by Hieron I
Theron of Acragas and Gelon of Syracuse defeated Carthaginians in 480, Destroyed by Carthage 409
Appealed (against Selinus) to Athens in 426 and 416 and Carthage in 410
Pan-Hellenic foundation under Athenian auspices (443)
Thurii
Messana
Rhegium
Panormus
Solus
Himera
Motya
Segesta
CARTHAGINIANS
Selinus
Mt. Etna
Naxos
Symaethus
Aetna
Catana
GREEKS
SICELS
Leontini
Palici
Acragas
Anapus
Syracuse
Gela
Camarina
Carthage
Renewed alliance with Athens 433
Supported Athens
Abortive national movement of Ducetius (466-450)
Capital of Ducetius
Fell to Carthage 409
Fell to Carthage 405
Supported Syracuse against Athens. Sacked by Carthage 405
Third foundation by Gela. Supported Athens (424) then Syracuse (413)
Empire of Hieron I (478-466). Democracy destroys Athenian invaders 413

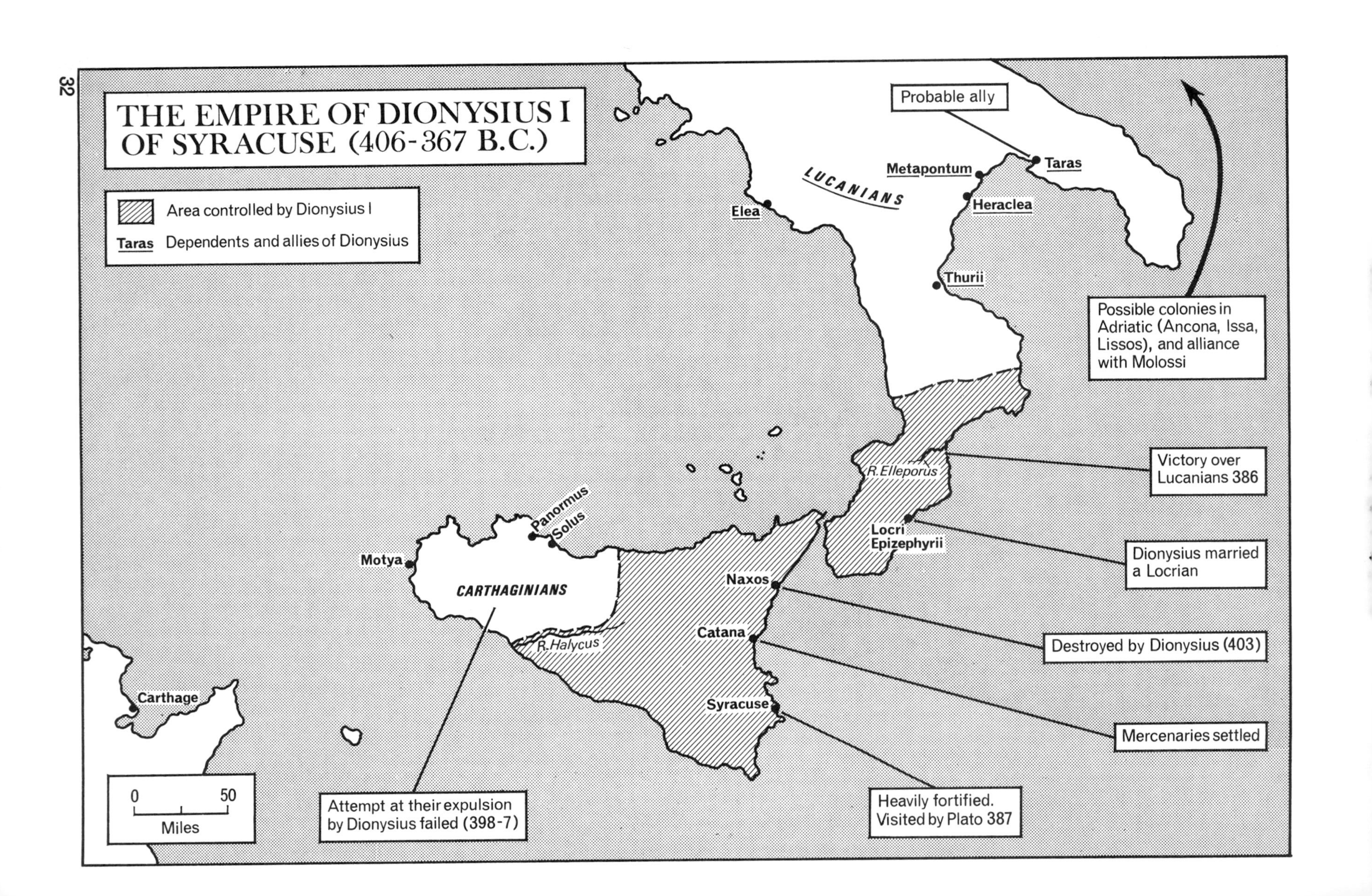
THE EMPIRE OF DIONYSIUS I OF SYRACUSE (406-367 B.C.)
Area controlled by Dionysius I
Taras Dependents and allies of Dionysius
Probable ally
Taras
Metapontum
Heraclea
Thurii
LUCANIANS
Elea
Possible colonies in Adriatic (Ancona, Issa, Lissos), and alliance with Molossi
R. Elleporus
Victory over Lucanians 386
Locri Epizephyrii
Dionysius married a Locrian
Panormus
Solus
Motya
CARTHAGINIANS
Naxos
Destroyed by Dionysius (403)
Catana
Mercenaries settled
R. Halycus
Syracuse
Heavily fortified. Visited by Plato 387
Carthage
Attempt at their expulsion by Dionysius failed (398-7)
0
50
Miles

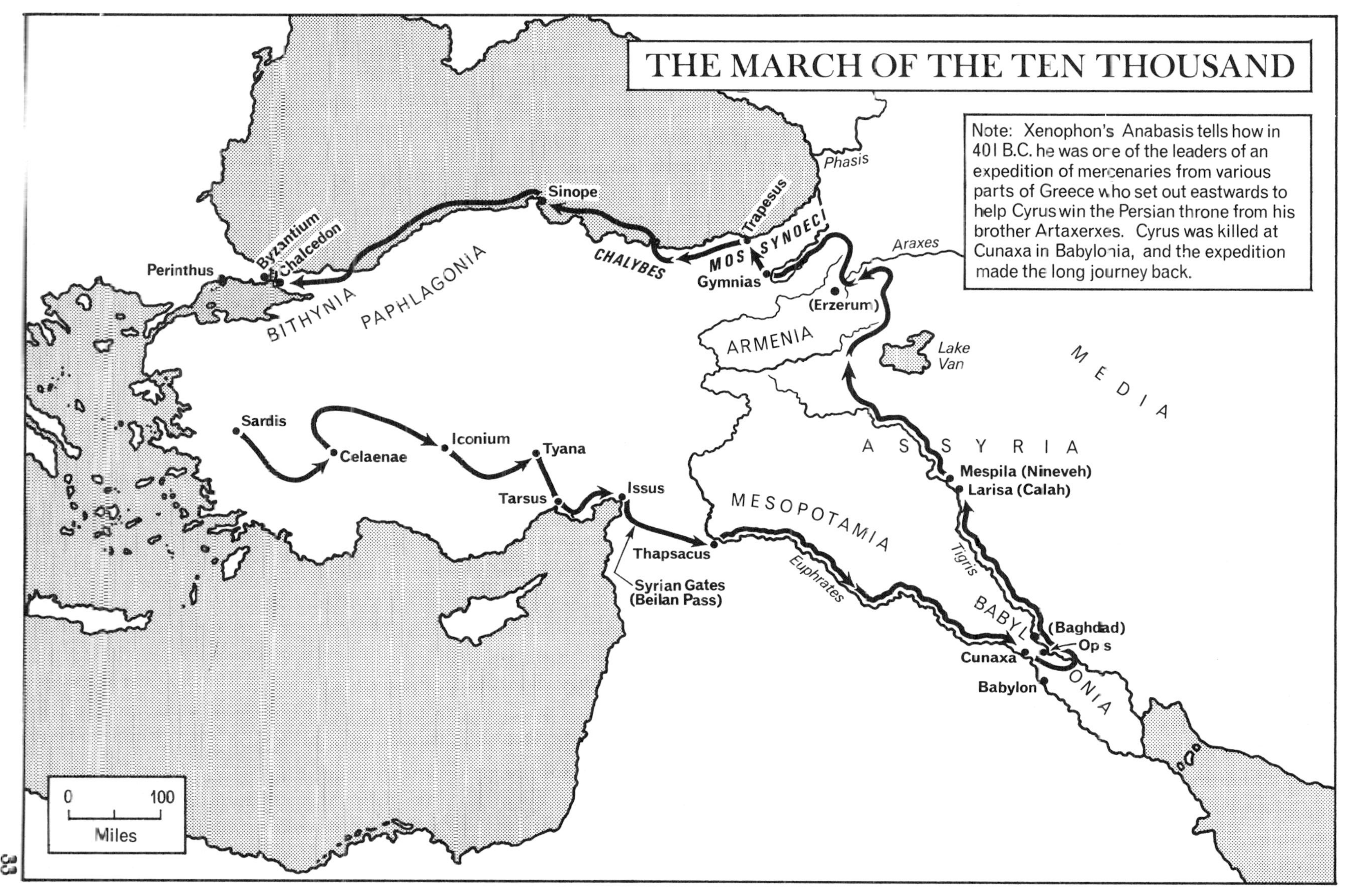
THE MARCH OF THE TEN THOUSAND
Note: Xenophon's Anabasis tells how in 401 B.C. he was one of the leaders of an expedition of mercenaries from various parts of Greece who set out eastwards to help Cyrus win the Persian throne from his brother Artaxerxes. Cyrus was killed at Cunaxa in Babylonia, and the expedition made the long journey back.
Phasis
Araxes
Sinope
Trapesus
SYNOECI
MOS
CHALYBES
Gymnias
Byzantium
Chalcedon
Perinthus
BITHYNIA
PAPHLAGONIA
(Erzerum)
ARMENIA
Lake Van
MEDIA
Sardis
Celaenae
Iconium
Tyana
Tarsus
Issus
ASSYRIA
Mespila (Nineveh)
Larisa (Calah)
MESOPOTAMIA
Thapsacus
Syrian Gates (Beilan Pass)
Euphrates
Tigris
BABYLONIA
(Baghdad)
Opis
Cunaxa
Babylon
0
100
Miles

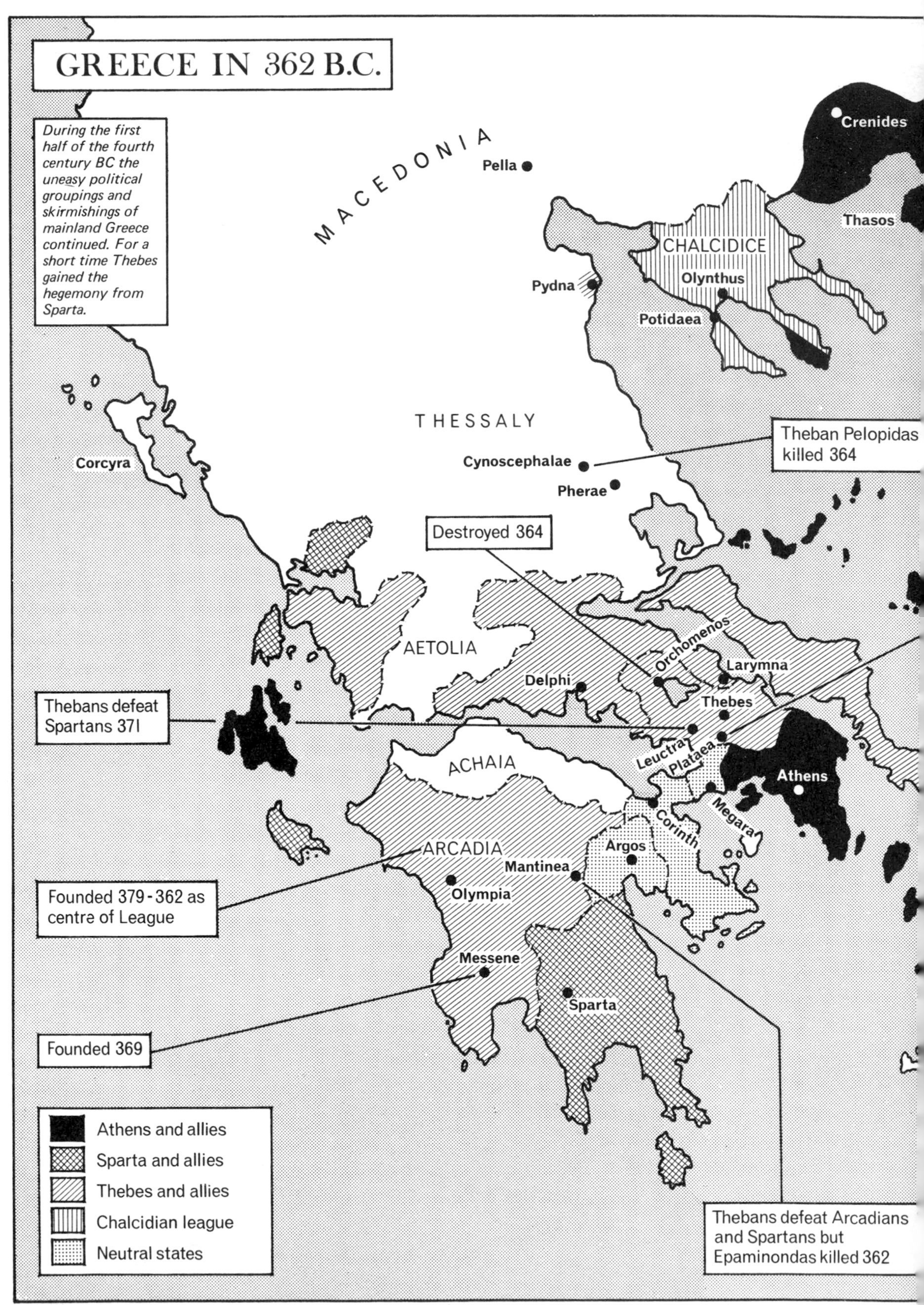
GREECE IN 362 B.C.
During the first half of the fourth century BC the uneasy political groupings and skirmishings of mainland Greece continued. For a short time Thebes gained the hegemony from Sparta.
MACEDONIA
Pella
Crenides
Thasos
CHALCIDICE
Olynthus
Pydna
Potidaea
THESSALY
Corcyra
Cynoscephalae
Pherae
Theban Pelopidas killed 364
Destroyed 364
AETOLIA
Orchomenos
Larymna
Delphi
Thebes
Thebans defeat Spartans 371
Leuctra
Plataea
ACHAIA
Athens
Corinth
Megara
ARCADIA
Argos
Mantinea
Olympia
Founded 379 - 362 as centre of League
Messene
Sparta
Founded 369
Athens and allies
Sparta and allies
Thebes and allies
Chalcidian league
Neutral states
Thebans defeat Arcadians and Spartans but Epaminondas killed 362

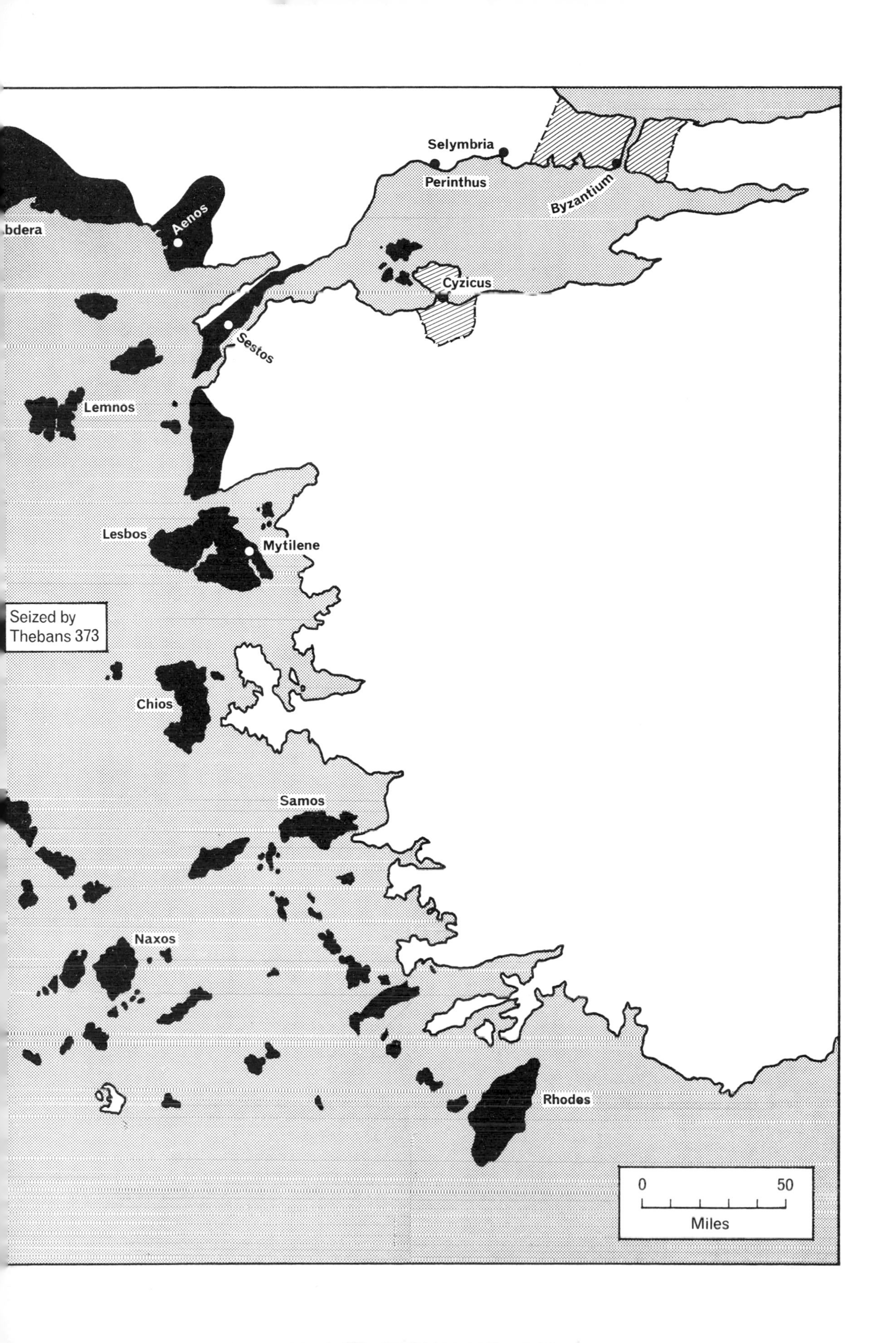
Selymbria
Perinthus
Byzantium
Aenos
bdera
Cyzicus
Sestos
Lemnos
Lesbos
Mytilene
Seized by
Thebans 373
Chios
Samos
Naxos
Rhodes
0
50
Miles

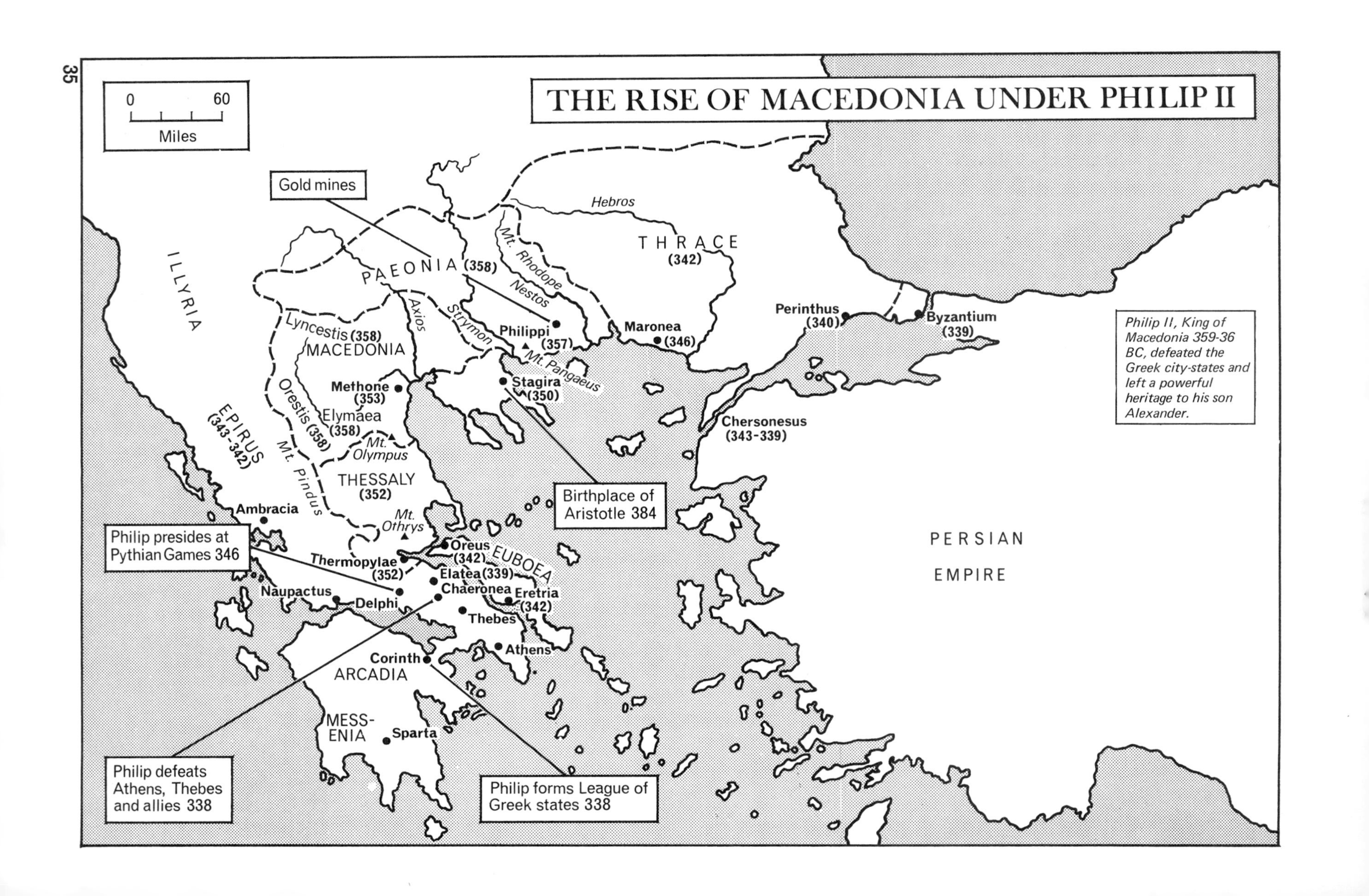
THE RISE OF MACEDONIA UNDER PHILIP II
0
60
Miles
Gold mines
ILLYRIA
PAEONIA (358)
Hebros
THRACE
(342)
Mt. Rhodope
Nestos
Axios
Strymon
Lyncestis (358)
MACEDONIA
Philippi
(357)
Mt. Pangaeus
Maronea
(346)
Perinthus
(340)
Byzantium
(339)
Philip II, King of Macedonia 359-36 BC, defeated the Greek city-states and left a powerful heritage to his son Alexander.
Methone
(353)
Stagira
(350)
Orestis (358)
Elymaea
(358)
EPIRUS
(343-342)
Mt. Olympus
Chersonesus
(343-339)
Mt. Pindus
THESSALY
(352)
Ambracia
Birthplace of Aristotle 384
Mt. Othrys
Philip presides at Pythian Games 346
Oreus
(342)
EUBOEA
Thermopylae
(352)
Elatea (339)
PERSIAN
EMPIRE
Naupactus
Delphi
Chaeronea
Eretria
(342)
Thebes
Athens
Corinth
ARCADIA
MESS-
ENIA
Sparta
Philip defeats Athens, Thebes and allies 338
Philip forms League of Greek states 338

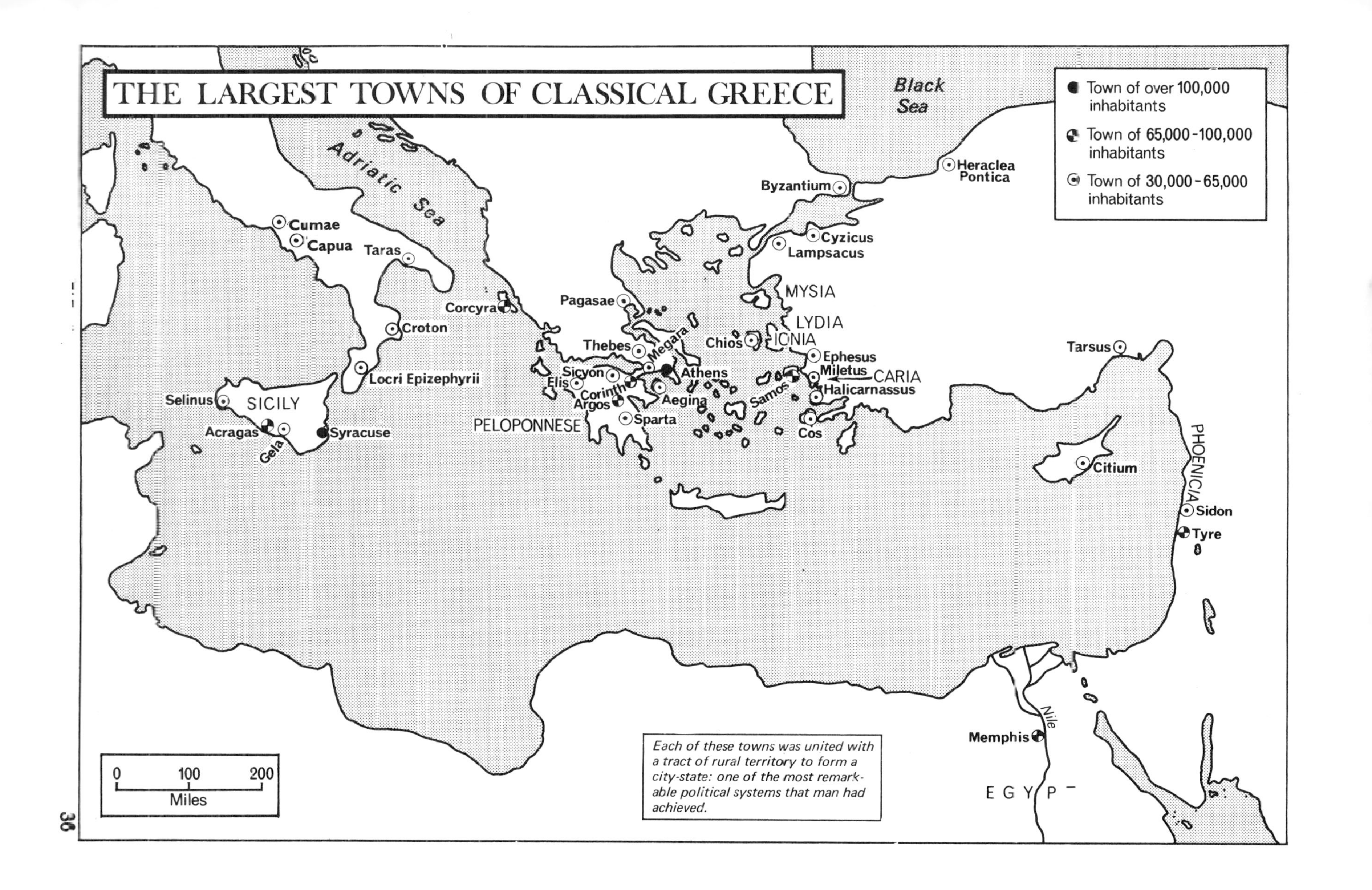
THE LARGEST TOWNS OF CLASSICAL GREECE
Town of over 100,000 inhabitants
Town of 65,000-100,000 inhabitants
Town of 30,000-65,000 inhabitants
Black Sea
Adriatic Sea
Heraclea Pontica
Byzantium
Cyzicus
Lampsacus
MYSIA
LYDIA
IONIA
CARIA
Chios
Ephesus
Miletus
Halicarnassus
Samos
Cos
Cumae
Capua
Taras
Corcyra
Croton
Locri Epizephyrii
Pagasae
Thebes
Megara
Athens
Sicyon
Elis
Corinth
Argos
Aegina
Sparta
PELOPONNESE
Selinus
SICILY
Acragas
Gela
Syracuse
Tarsus
Citium
PHOENICIA
Sidon
Tyre
Memphis
Nile
EGYPT
0 100 200
Miles
Each of these towns was united with a tract of rural territory to form a city-state: one of the most remarkable political systems that man had achieved.

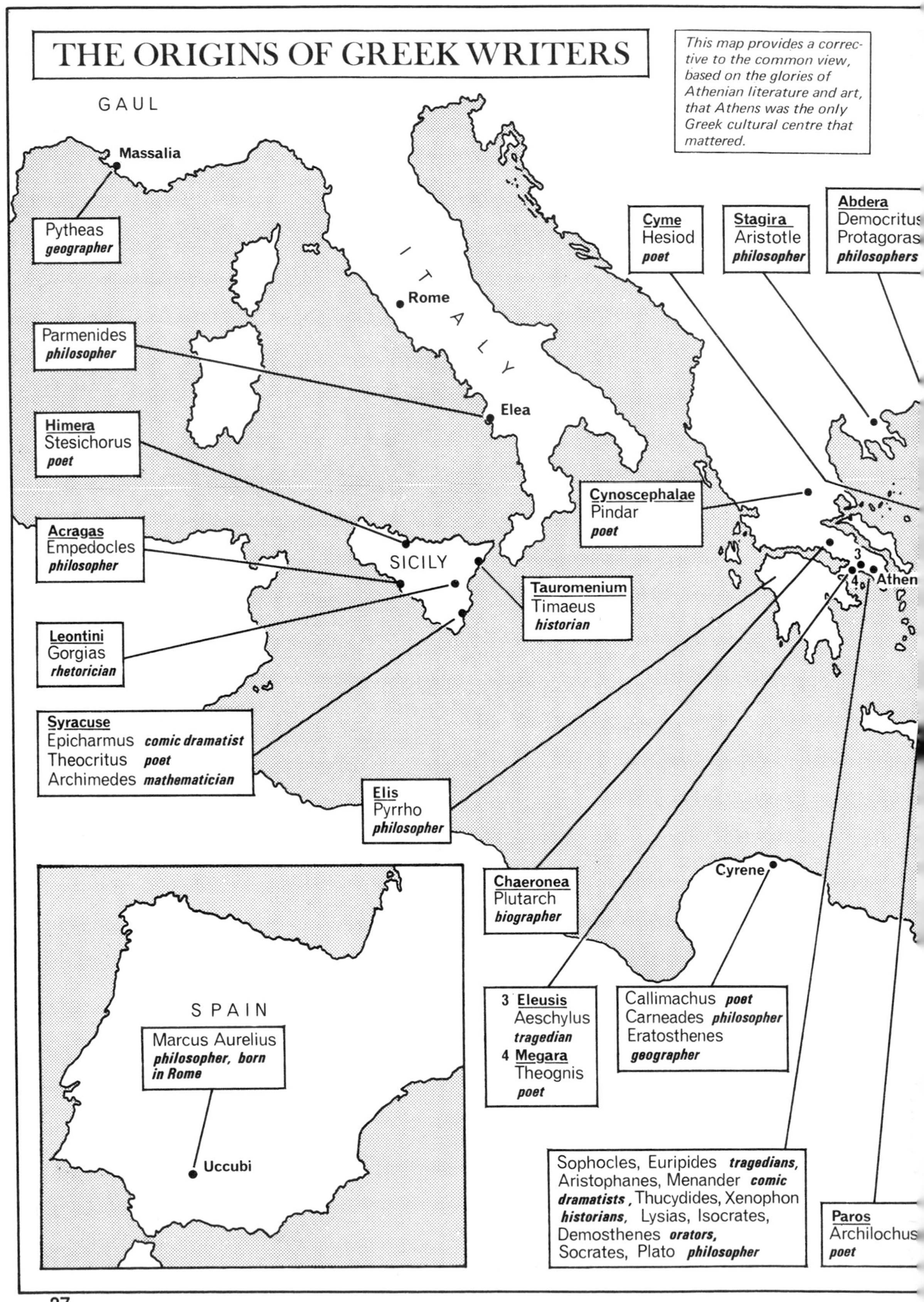
THE ORIGINS OF GREEK WRITERS
This map provides a corrective to the common view, based on the glories of Athenian literature and art, that Athens was the only Greek cultural centre that mattered.
GAUL
Massalia
Pytheas geographer
ITALY
Rome
Parmenides philosopher
Elea
Himera Stesichorus poet
Acragas Empedocles philosopher
SICILY
Leontini Gorgias rhetorician
Syracuse
Epicharmus comic dramatist
Theocritus poet
Archimedes mathematician
Tauromenium Timaeus historian
Elis Pyrrho philosopher
Cynoscephalae Pindar poet
Cyme Hesiod poet
Stagira Aristotle philosopher
Abdera Democritus Protagoras philosophers
Athen
Chaeronea Plutarch biographer
Cyrene
3 Eleusis Aeschylus tragedian
4 Megara Theognis poet
Callimachus poet
Carneades philosopher
Eratosthenes geographer
Sophocles, Euripides tragedians, Aristophanes, Menander comic dramatists, Thucydides, Xenophon historians, Lysias, Isocrates, Demosthenes orators, Socrates, Plato philosopher
Paros Archilochus poet
SPAIN
Marcus Aurelius philosopher, born in Rome
Uccubi

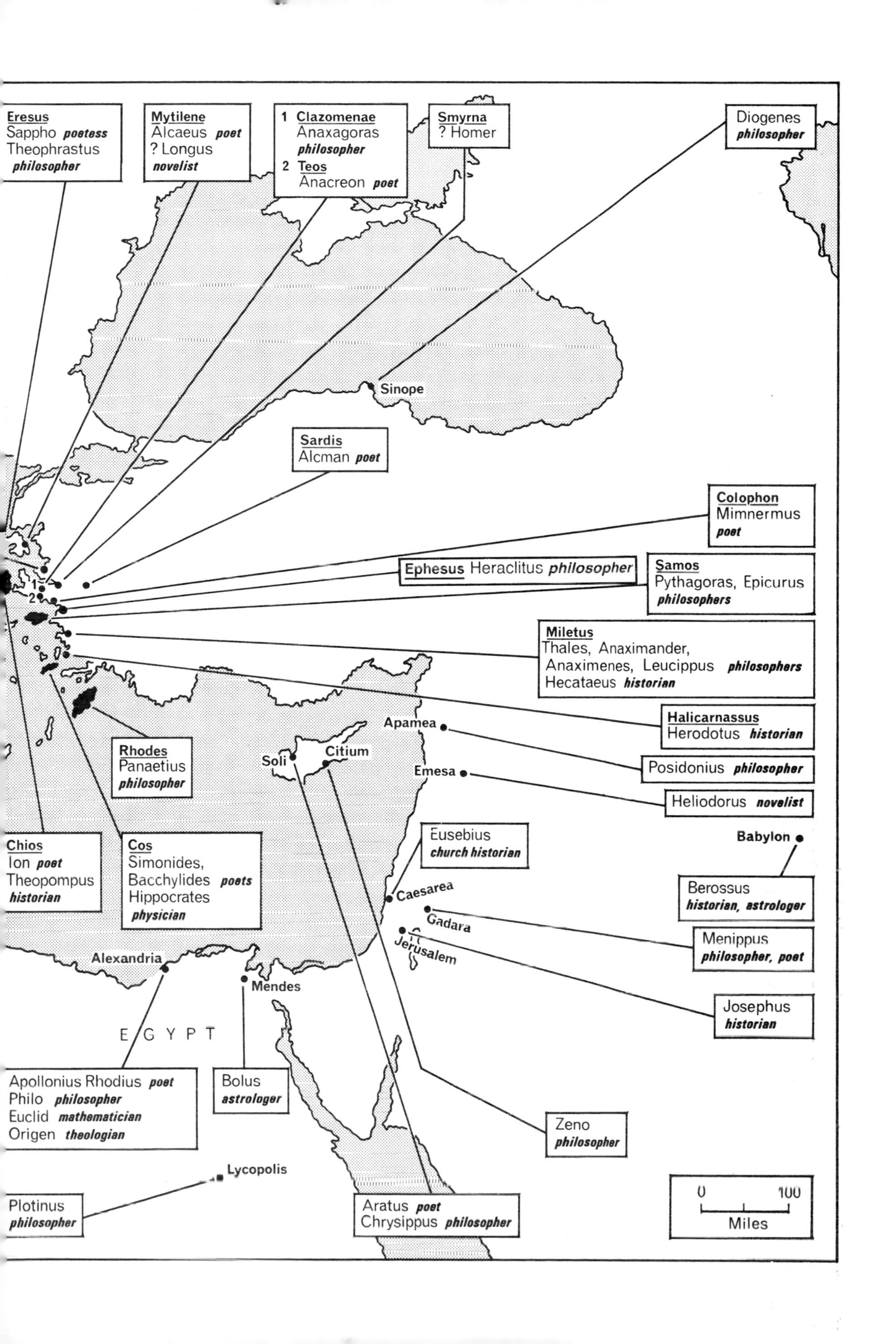

Eresus
Sappho poetess
Theophrastus philosopher
Mytilene
Alcaeus poet
? Longus novelist
1 Clazomenae
Anaxagoras philosopher
2 Teos
Anacreon poet
Smyrna
? Homer
Diogenes philosopher
Sinope
Sardis
Alcman poet
Colophon
Mimnermus poet
Ephesus Heraclitus philosopher
Samos
Pythagoras, Epicurus philosophers
Miletus
Thales, Anaximander, Anaximenes, Leucippus philosophers
Hecataeus historian
Halicarnassus
Herodotus historian
Apamea
Posidonius philosopher
Soli
Citium
Emesa
Heliodorus novelist
Rhodes
Panaetius philosopher
Chios
Ion poet
Theopompus historian
Cos
Simonides, Bacchylides poets
Hippocrates physician
Eusebius church historian
Babylon
Berossus historian, astrologer
Caesarea
Gadara
Jerusalem
Menippus philosopher, poet
Josephus historian
Alexandria
Mendes
EGYPT
Apollonius Rhodius poet
Philo philosopher
Euclid mathematician
Origen theologian
Bolus astrologer
Zeno philosopher
Lycopolis
Plotinus philosopher
Aratus poet
Chrysippus philosopher
0 100
Miles

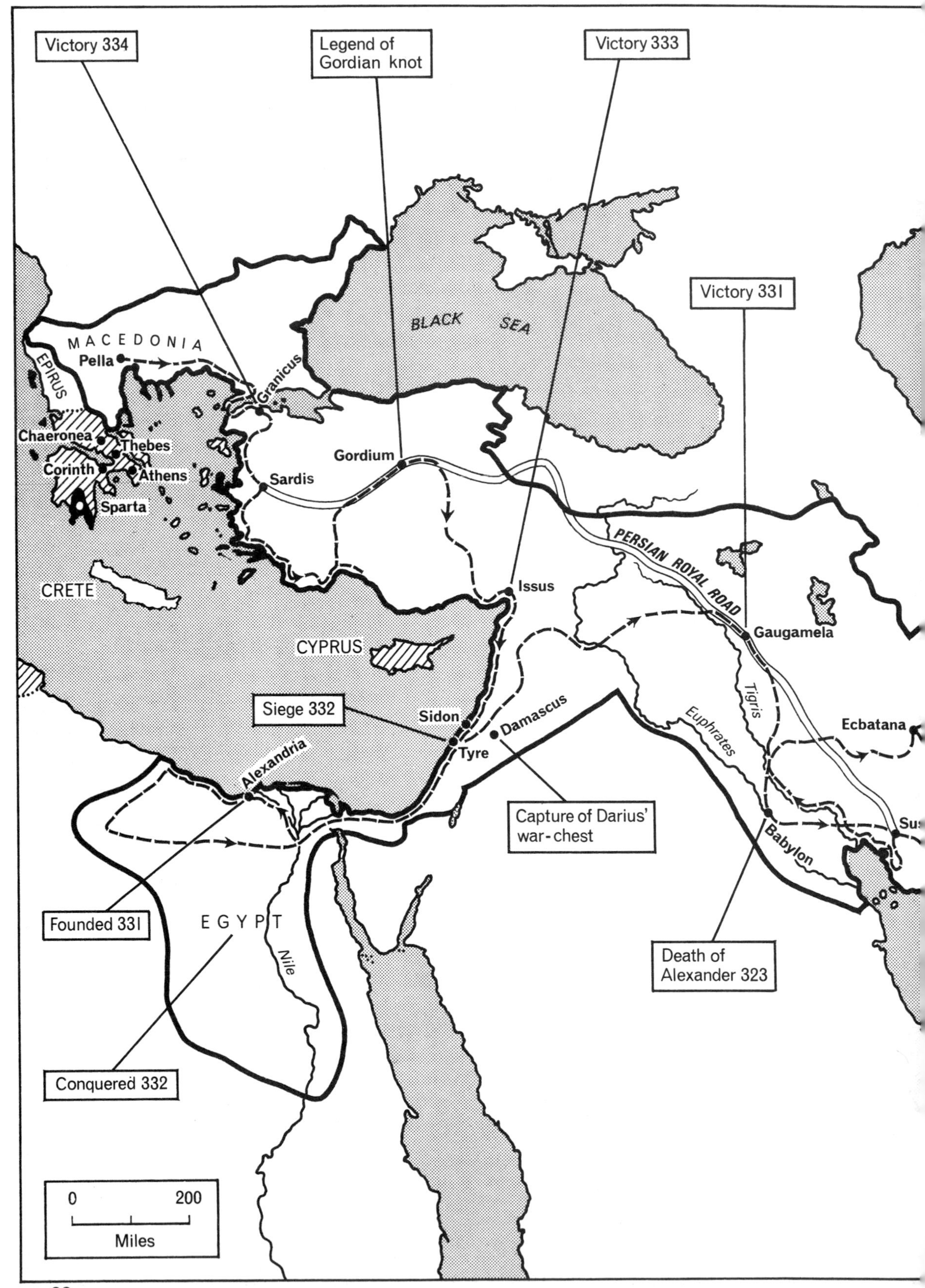
Victory 334
Legend of Gordian knot
Victory 333
Victory 331
BLACK SEA
MACEDONIA
EPIRUS
Pella
Granicus
Chaeronea
Thebes
Corinth
Athens
Sparta
Gordium
Sardis
PERSIAN ROYAL ROAD
CRETE
Issus
Gaugamela
CYPRUS
Tigris
Euphrates
Siege 332
Sidon
Damascus
Tyre
Ecbatana
Alexandria
Capture of Darius' war-chest
Babylon
Founded 331
EGYPT
Nile
Death of Alexander 323
Conquered 332
0
200
Miles

THE CONQUESTS OF ALEXANDER THE GREAT

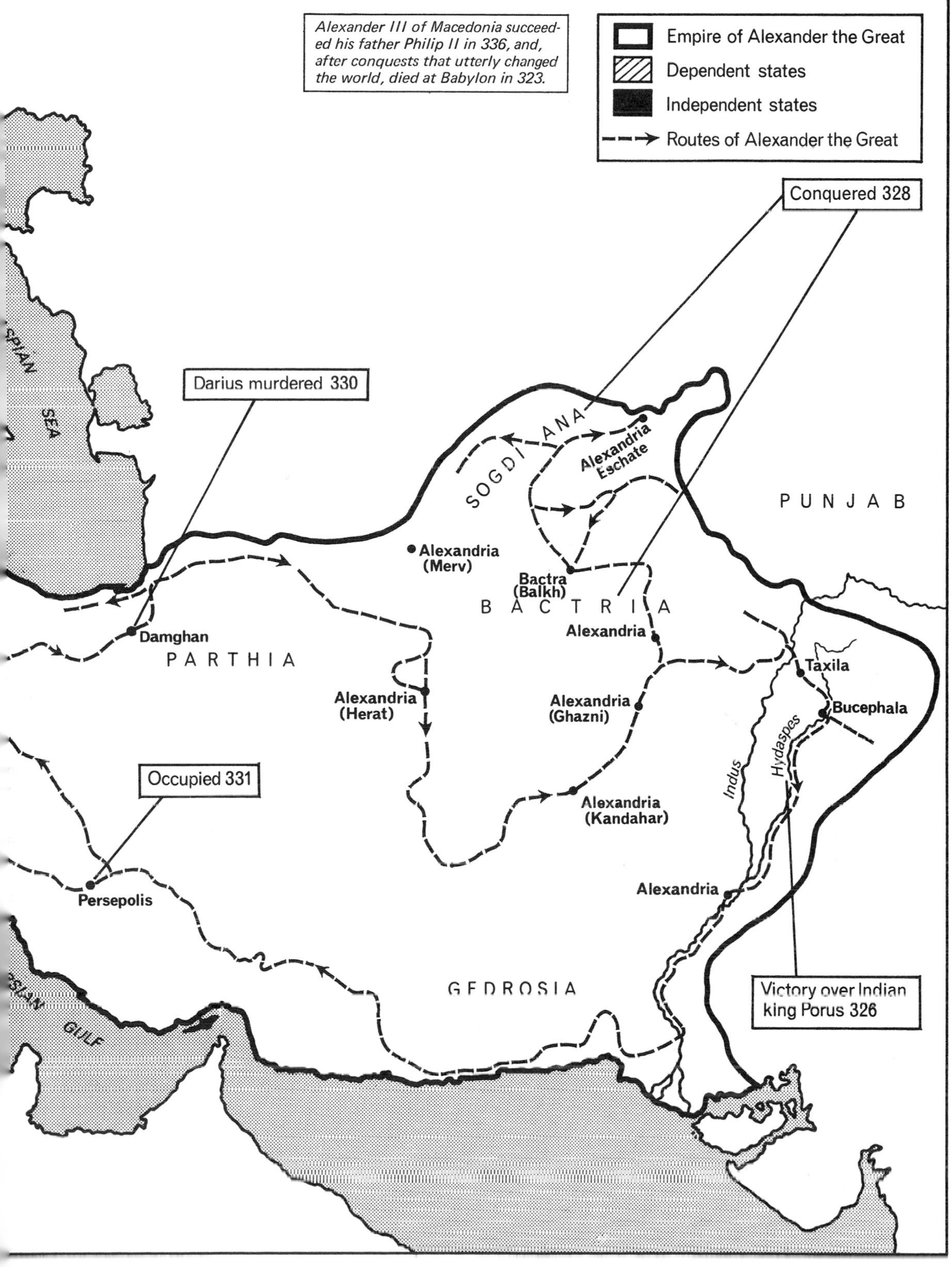

Alexander III of Macedonia succeeded his father Philip II in 336, and, after conquests that utterly changed the world, died at Babylon in 323.

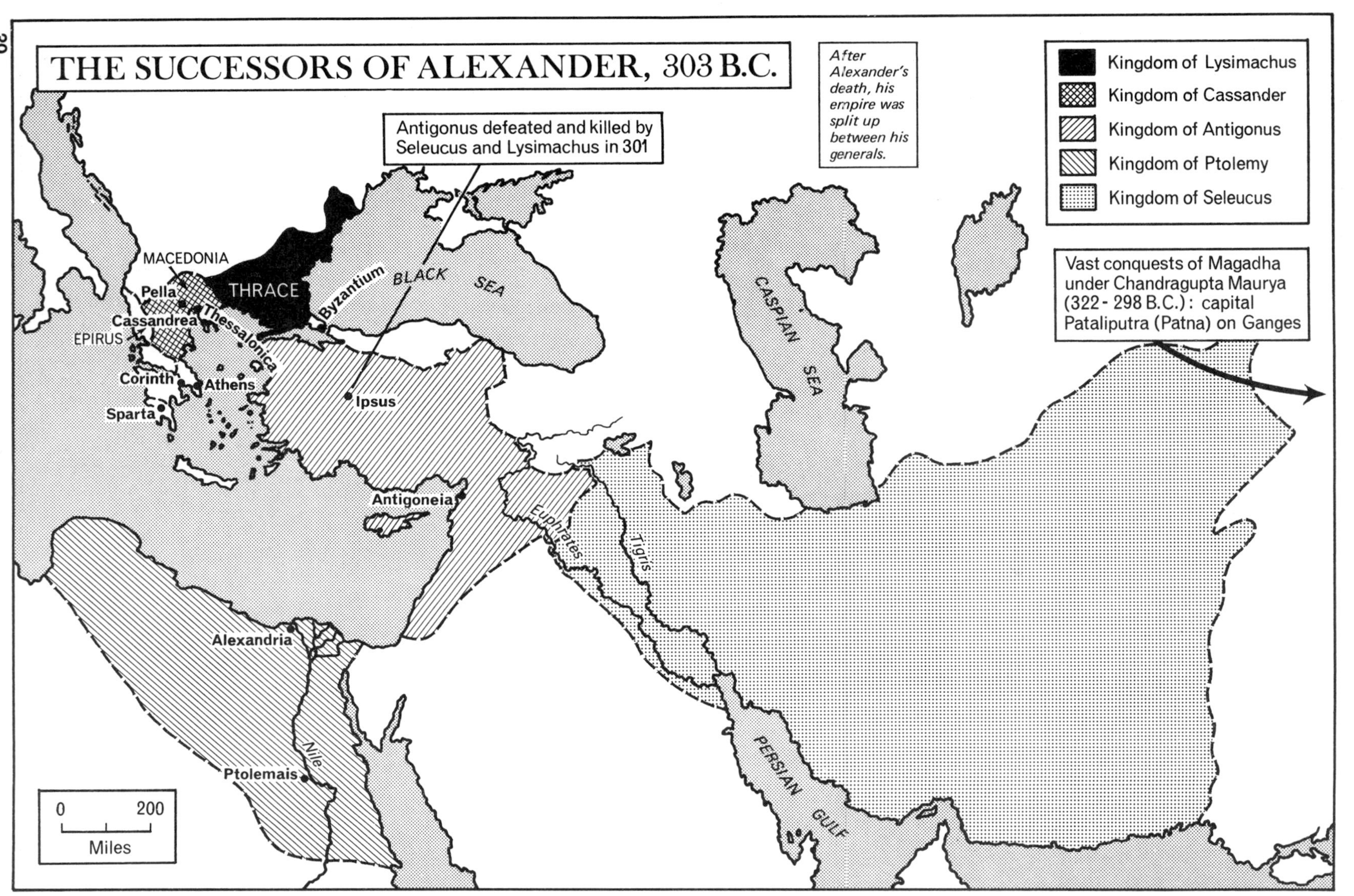

THE SUCCESSORS OF ALEXANDER, 303 B.C.
After Alexander's death, his empire was split up between his generals.
Kingdom of Lysimachus
Kingdom of Cassander
Kingdom of Antigonus
Kingdom of Ptolemy
Kingdom of Seleucus
Antigonus defeated and killed by Seleucus and Lysimachus in 301
Vast conquests of Magadha under Chandragupta Maurya (322 - 298 B.C.): capital Pataliputra (Patna) on Ganges
MACEDONIA
THRACE
Pella
Cassandrea
Thessalonica
EPIRUS
Corinth
Athens
Sparta
Byzantium
BLACK SEA
Ipsus
Antigoneia
CASPIAN SEA
Euphrates
Tigris
PERSIAN GULF
Alexandria
Nile
Ptolemais
0 200
Miles

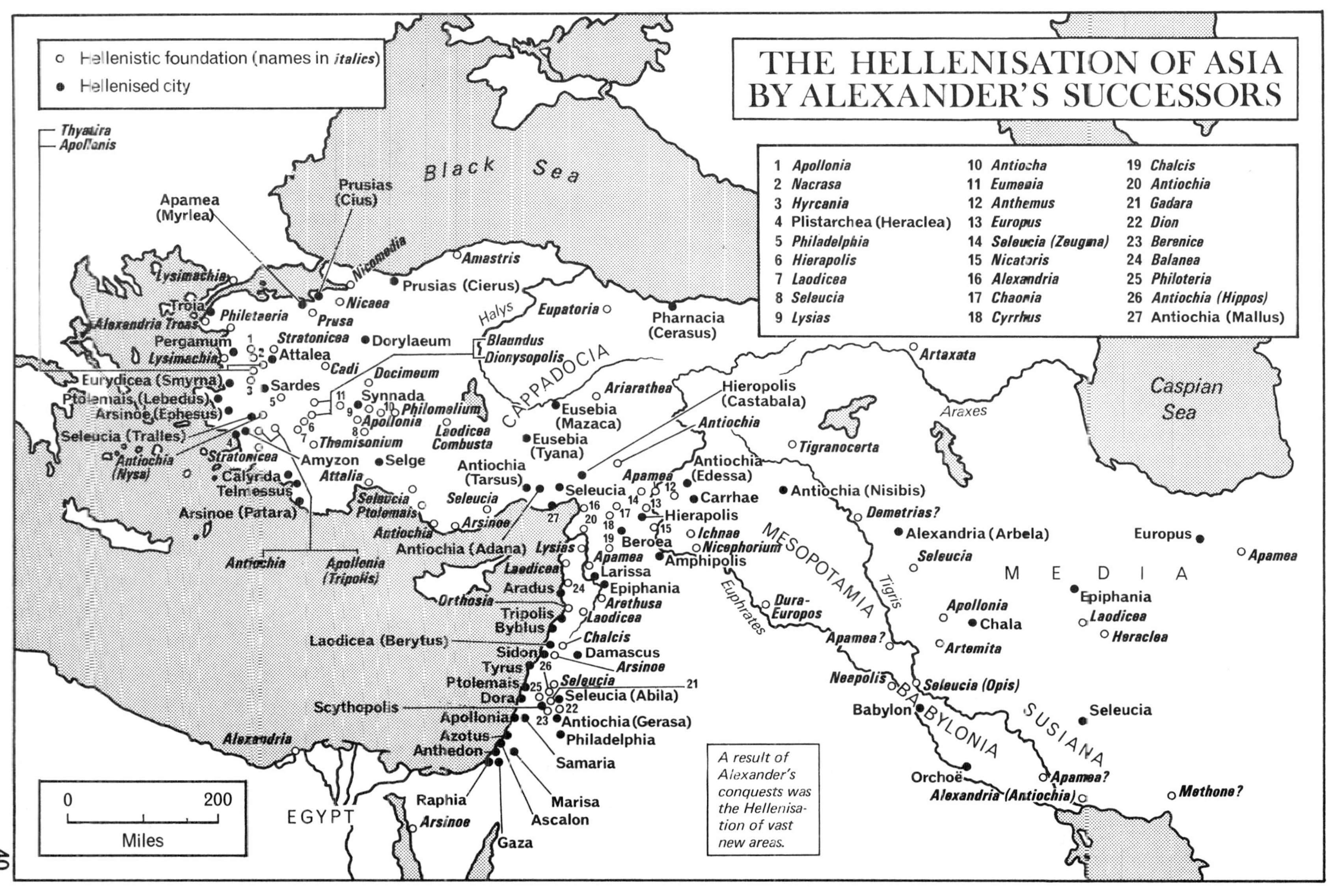
THE HELLENISATION OF ASIA
BY ALEXANDER'S SUCCESSORS
Hellenistic foundation (names in italics)
Hellenised city
1 Apollonia
2 Nacrasa
3 Hyrcania
4 Plistarchea (Heraclea)
5 Philadelphia
6 Hierapolis
7 Laodicea
8 Seleucia
9 Lysias
10 Antiocha
11 Eumeaia
12 Anthemus
13 Europus
14 Seleucia (Zeugma)
15 Nicatoris
16 Alexandria
17 Chaonia
18 Cyrrhus
19 Chalcis
20 Antiochia
21 Gadara
22 Dion
23 Berenice
24 Balanea
25 Philoteria
26 Antiochia (Hippos)
27 Antiochia (Mallus)
Black Sea
Caspian Sea
Thyatira
Apollonis
Apamea (Myrlea)
Prusias (Cius)
Nicomedia
Amastris
Prusias (Cierus)
Nicaea
Prusa
Lysimachia
Troja
Alexandria Troas
Philetaeria
Pergamum
Lysimachia
Stratonicea
Attalea
Dorylaeum
Halys
Eupatoria
Pharnacia (Cerasus)
Blaundus
Dionysopolis
CAPPADOCIA
Cadi
Docimeum
Eurydicea (Smyrna)
Ptolemais (Lebedus)
Arsinoe (Ephesus)
Sardes
Synnada
Philomelium
Apollonia
Themisonium
Laodicea Combusta
Seleucia (Tralles)
Antiochia (Nysa)
Stratonicea
Amyzon
Selge
Calyrda
Attalia
Telmessus
Arsinoe (Patara)
Seleucia
Ptolemais
Antiochia
Antiochia (Tarsus)
Seleucia
Arsinoe
Antiochia (Adana)
Antiochia
Apollonia (Tripolis)
Ariarathea
Eusebia (Mazaca)
Eusebia (Tyana)
Hieropolis (Castabala)
Antiochia
Artaxata
Araxes
Tigranocerta
Apamea
Antiochia (Edessa)
Carrhae
Hierapolis
Antiochia (Nisibis)
Demetrias?
Beroea
Ichnae
Nicephorium
Amphipolis
MESOPOTAMIA
Lysias
Apamea
Larissa
Laodicea
Aradus
Epiphania
Arethusa
Laodicea
Orthosia
Tripolis
Byblus
Chalcis
Laodicea (Berytus)
Sidon
Damascus
Arsinoe
Tyrus
Ptolemais
Seleucia
Dora
Seleucia (Abila)
Scythopolis
Apollonia
Antiochia (Gerasa)
Azotus
Philadelphia
Anthedon
Samaria
Alexandria
EGYPT
Raphia
Arsinoe
Marisa
Ascalon
Gaza
Euphrates
Dura-Europos
Tigris
Apamea?
Alexandria (Arbela)
Seleucia
Europus
Apamea
MEDIA
Epiphania
Laodicea
Heraclea
Apollonia
Chala
Artemita
Neapolis
Seleucia (Opis)
Babylon
BABYLONIA
SUSIANA
Seleucia
Orchoë
Apamea?
Alexandria (Antiochia)
Methone?
0 200 Miles
A result of Alexander's conquests was the Hellenisation of vast new areas.

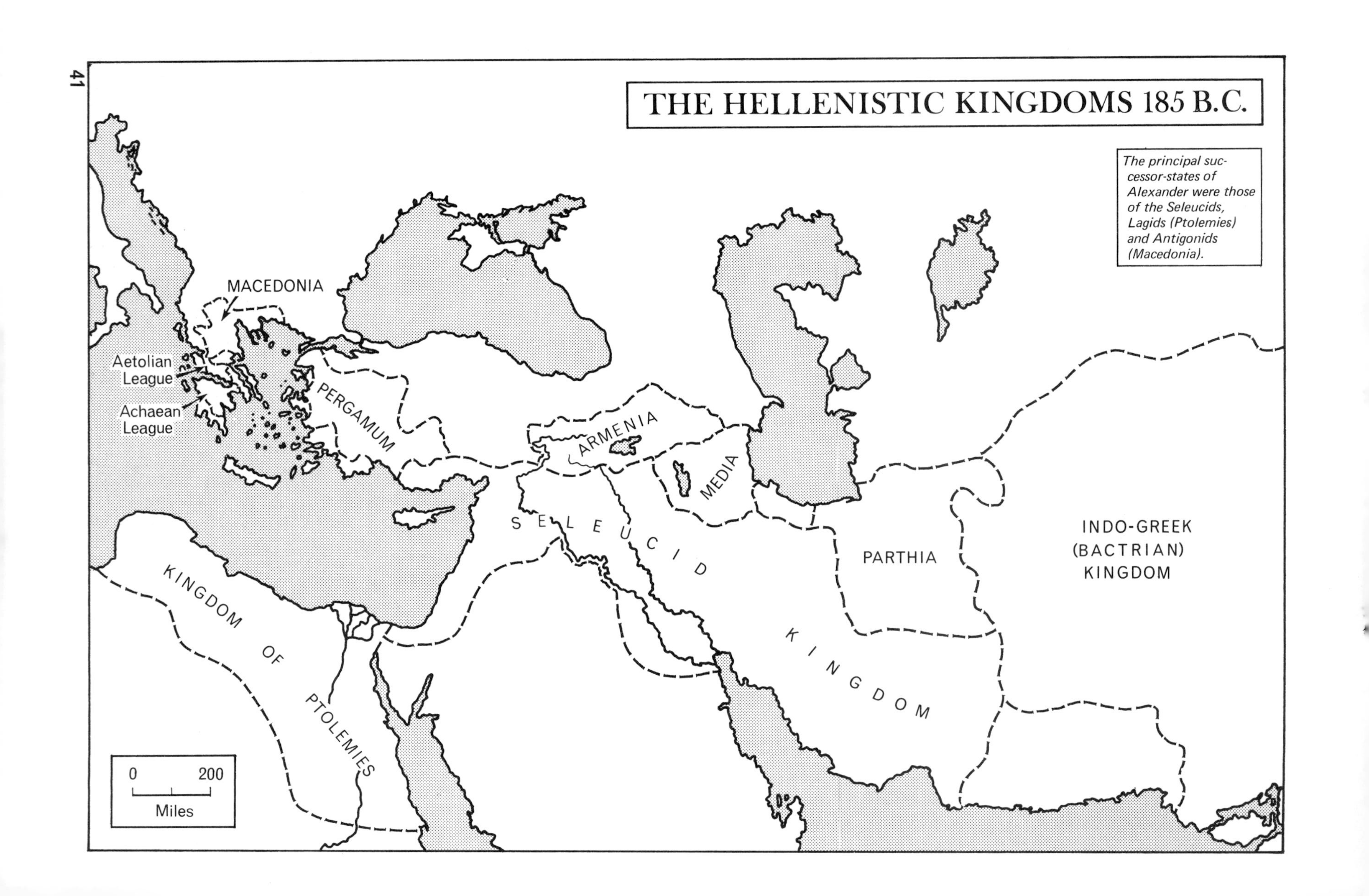

THE HELLENISTIC KINGDOMS 185 B.C.
The principal successor-states of Alexander were those of the Seleucids, Lagids (Ptolemies) and Antigonids (Macedonia).
MACEDONIA
Aetolian League
Achaean League
PERGAMUM
ARMENIA
MEDIA
SELEUCID KINGDOM
PARTHIA
INDO-GREEK (BACTRIAN) KINGDOM
KINGDOM OF PTOLEMIES
0
200
Miles

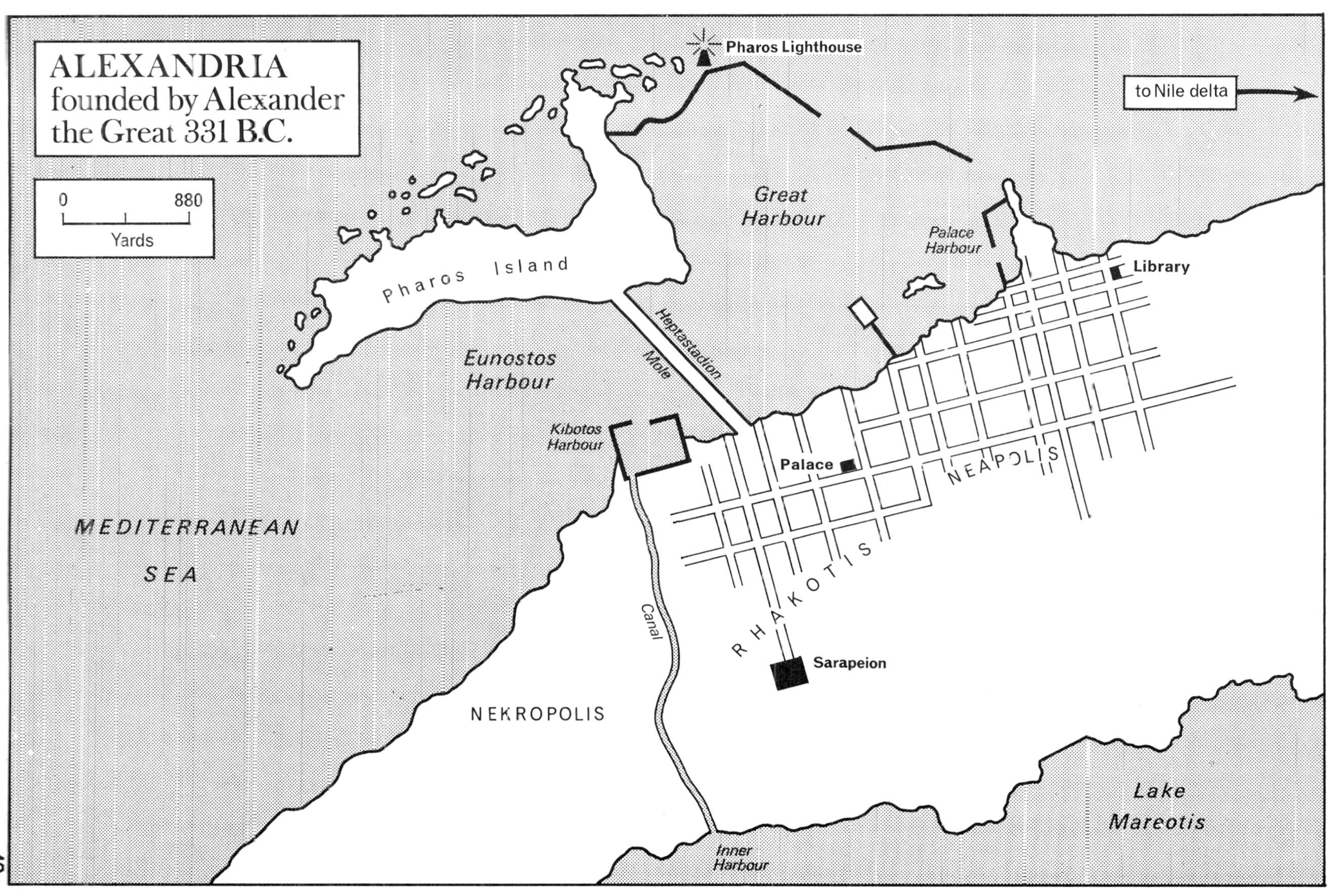

ALEXANDRIA
founded by Alexander
the Great 331 B.C.
0
880
Yards
Pharos Lighthouse
to Nile delta
Great
Harbour
Palace
Harbour
Library
Pharos Island
Heptastadion
Mole
Eunostos
Harbour
Kibotos
Harbour
Palace
NEAPOLIS
RHAKOTIS
Sarapeion
MEDITERRANEAN
SEA
Canal
NEKROPOLIS
Lake
Mareotis
Inner
Harbour

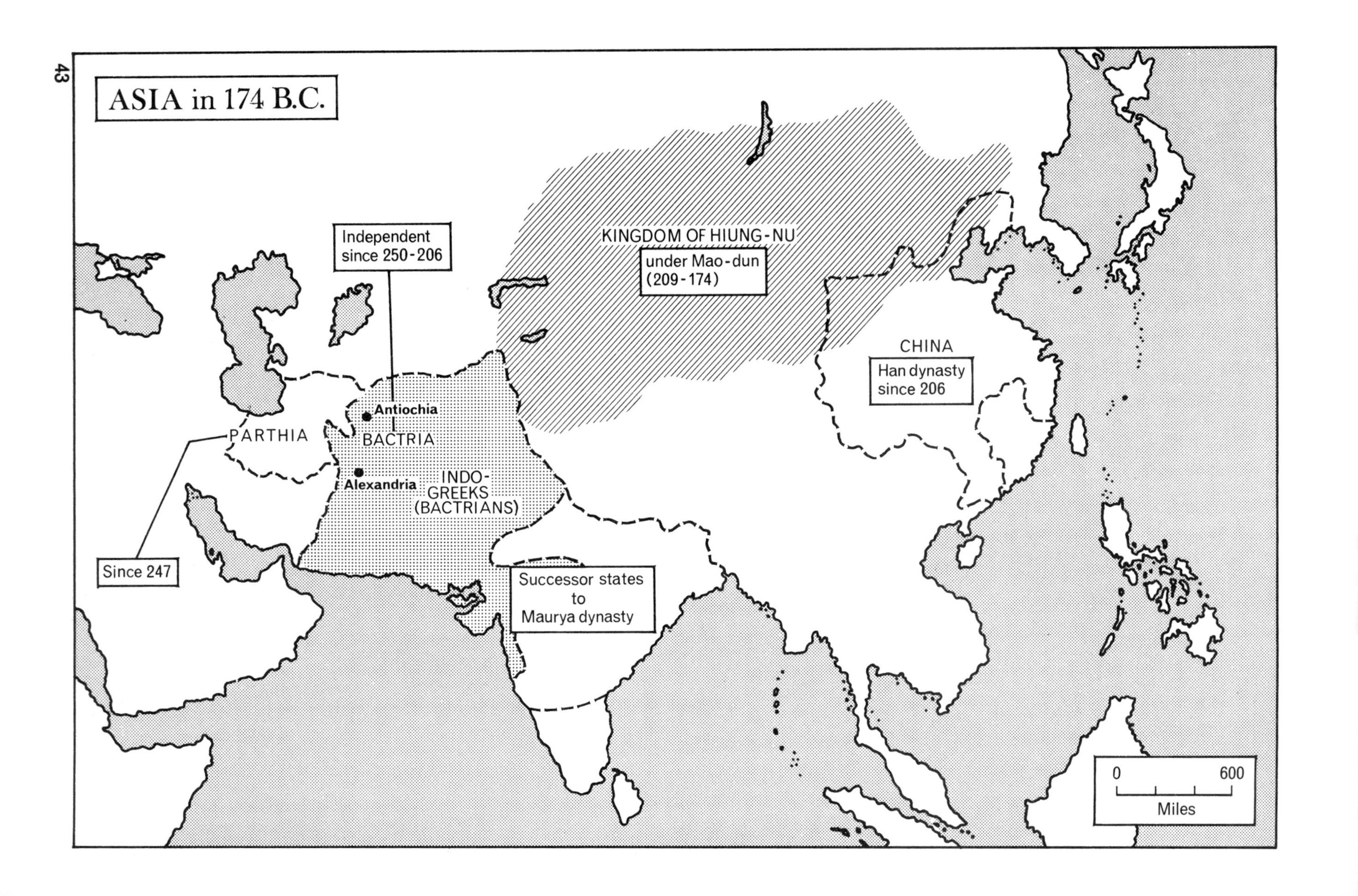
ASIA in 174 B.C.
Independent since 250-206
KINGDOM OF HIUNG-NU
under Mao-dun (209-174)
CHINA
Han dynasty since 206
Antiochia
BACTRIA
PARTHIA
Since 247
Alexandria
INDO-GREEKS (BACTRIANS)
Successor states to Maurya dynasty
0
600
Miles

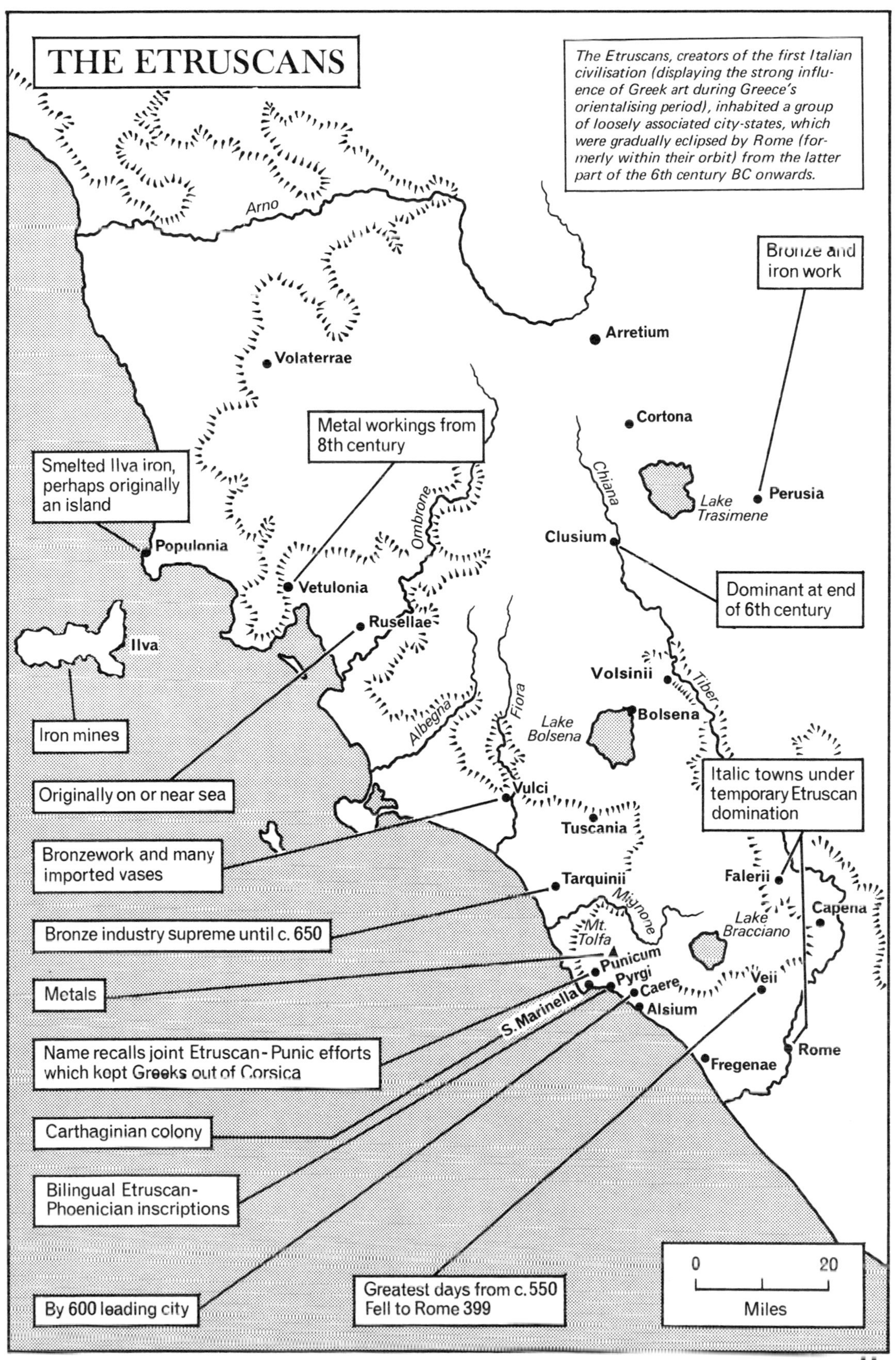
THE ETRUSCANS
The Etruscans, creators of the first Italian civilisation (displaying the strong influence of Greek art during Greece's orientalising period), inhabited a group of loosely associated city-states, which were gradually eclipsed by Rome (formerly within their orbit) from the latter part of the 6th century BC onwards.
Arno
Bronze and iron work
Arretium
Volaterrae
Cortona
Metal workings from 8th century
Chiana
Smelted Ilva iron, perhaps originally an island
Lake Trasimene
Perusia
Ombrone
Clusium
Populonia
Vetulonia
Dominant at end of 6th century
Rusellae
Ilva
Volsinii
Tiber
Albegna
Fiora
Lake Bolsena
Bolsena
Iron mines
Italic towns under temporary Etruscan domination
Originally on or near sea
Vulci
Tuscania
Bronzework and many imported vases
Tarquinii
Falerii
Mignone
Capena
Bronze industry supreme until c. 650
Mt. Tolfa
Lake Bracciano
Metals
Punicum
Pyrgi
Caere
Veii
S. Marinella
Alsium
Name recalls joint Etruscan-Punic efforts which kept Greeks out of Corsica
Rome
Fregenae
Carthaginian colony
Bilingual Etruscan-Phoenician inscriptions
0
20
Miles
By 600 leading city
Greatest days from c. 550 Fell to Rome 399

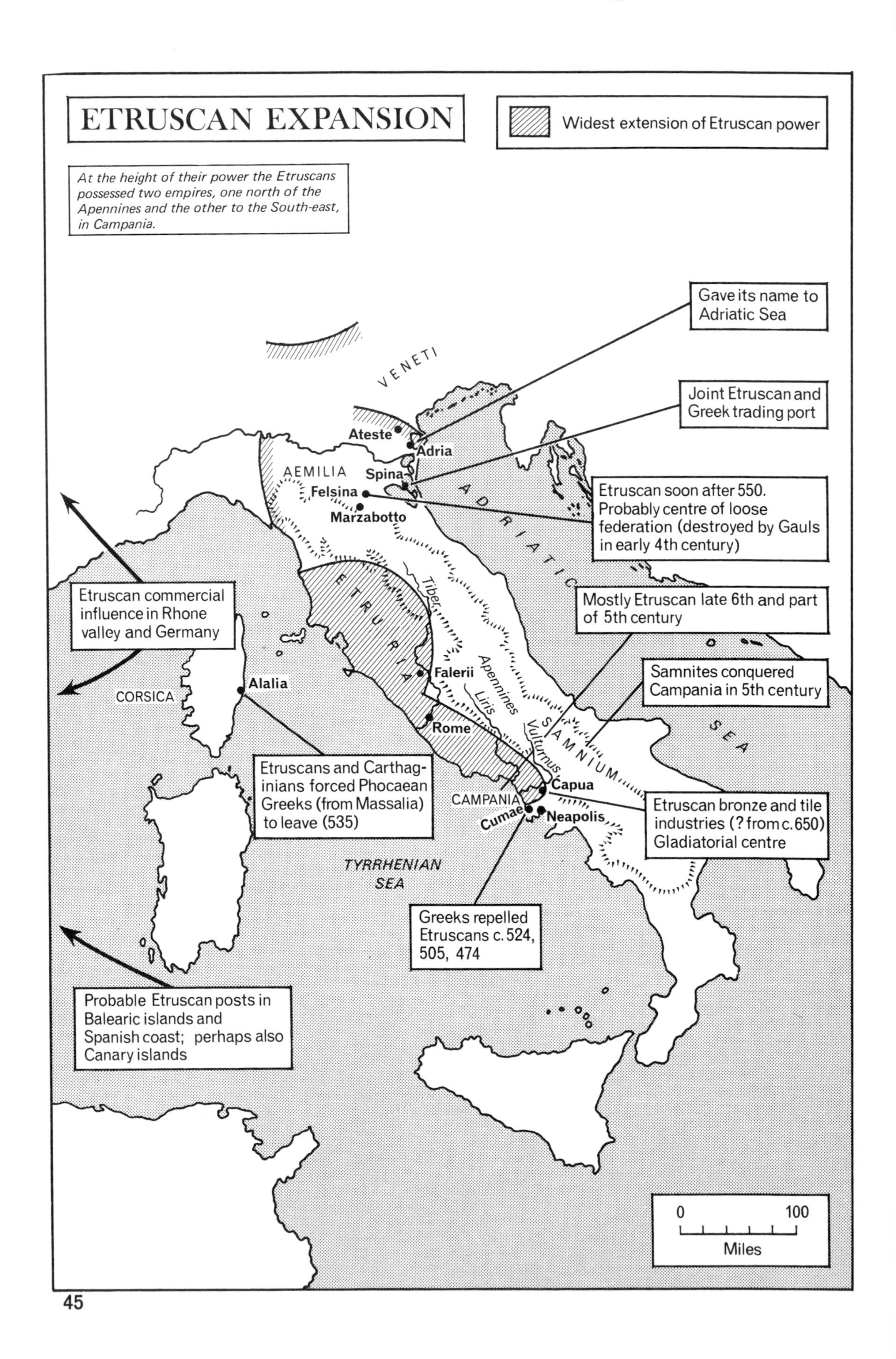
ETRUSCAN EXPANSION
Widest extension of Etruscan power
At the height of their power the Etruscans possessed two empires, one north of the Apennines and the other to the South-east, in Campania.
Gave its name to Adriatic Sea
Joint Etruscan and Greek trading port
Etruscan soon after 550. Probably centre of loose federation (destroyed by Gauls in early 4th century)
Mostly Etruscan late 6th and part of 5th century
Samnites conquered Campania in 5th century
Etruscan bronze and tile industries (? from c. 650) Gladiatorial centre
Greeks repelled Etruscans c. 524, 505, 474
Etruscans and Carthaginians forced Phocaean Greeks (from Massalia) to leave (535)
Etruscan commercial influence in Rhone valley and Germany
Probable Etruscan posts in Balearic islands and Spanish coast; perhaps also Canary islands
VENETI
Ateste
Adria
Spina
AEMILIA
Felsina
Marzabotto
ADRIATIC SEA
ETRURIA
Tiber
Apennines
Liris
Vulturnus
Falerii
Rome
SAMNIUM
Capua
CAMPANIA
Cumae
Neapolis
CORSICA
Alalia
TYRRHENIAN SEA
0
100
Miles

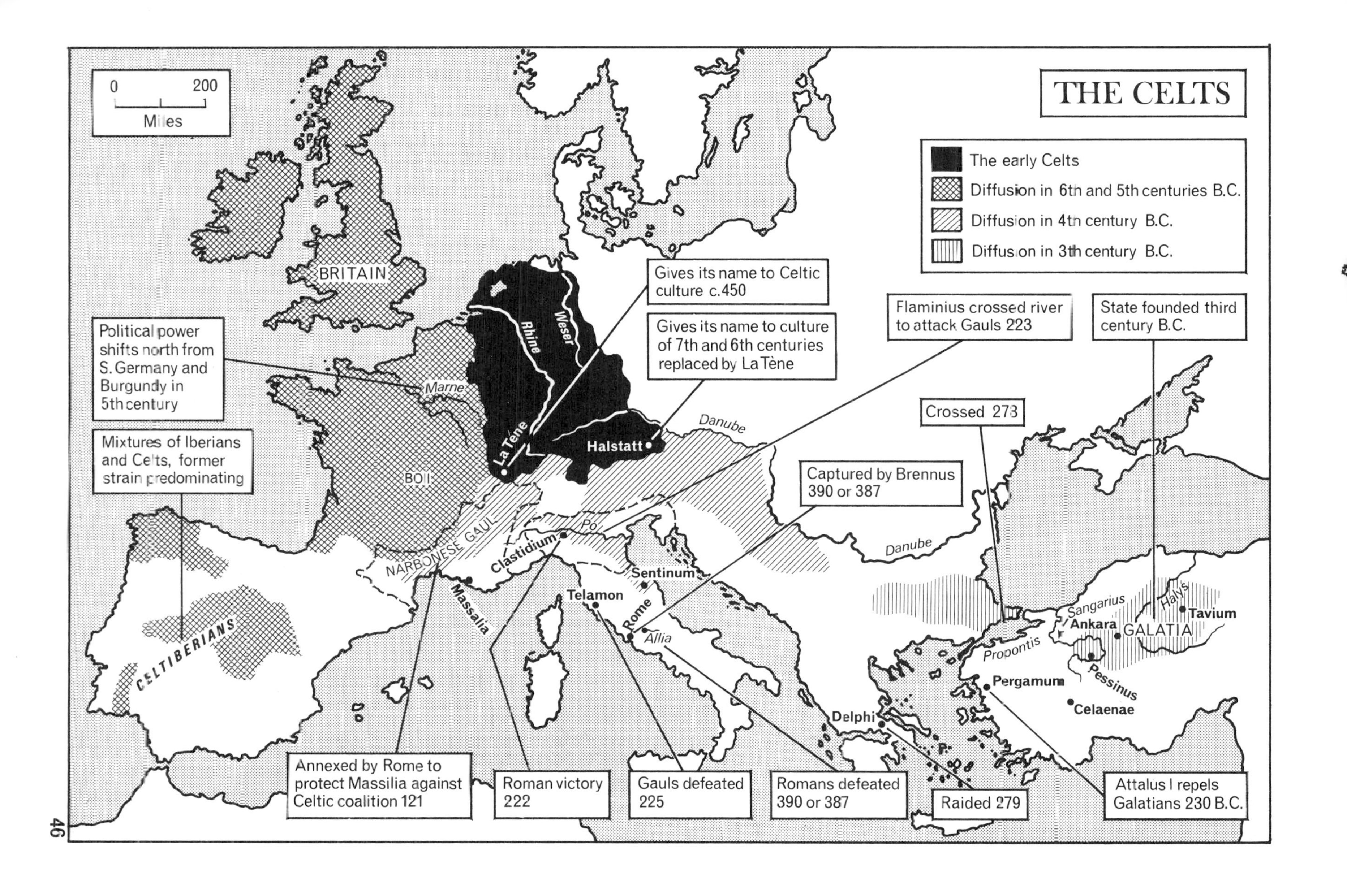
THE CELTS
0 200
Miles
The early Celts
Diffusion in 6th and 5th centuries B.C.
Diffusion in 4th century B.C.
Diffusion in 3th century B.C.
BRITAIN
Political power shifts north from S. Germany and Burgundy in 5th century
Mixtures of Iberians and Celts, former strain predominating
Gives its name to Celtic culture c.450
Gives its name to culture of 7th and 6th centuries replaced by La Tène
Flaminius crossed river to attack Gauls 223
State founded third century B.C.
Crossed 273
Captured by Brennus 390 or 387
Annexed by Rome to protect Massilia against Celtic coalition 121
Roman victory 222
Gauls defeated 225
Romans defeated 390 or 387
Raided 279
Attalus I repels Galatians 230 B.C.
Rhine
Weser
Marne
La Tène
Halstatt
Danube
BOII
NARBONESE GAUL
Po
Clastidium
Massalia
Telamon
Sentinum
Rome
Allia
CELTIBERIANS
Delphi
Propontis
Sangarius
Halys
Ankara
GALATIA
Tavium
Pessinus
Pergamum
Celaenae

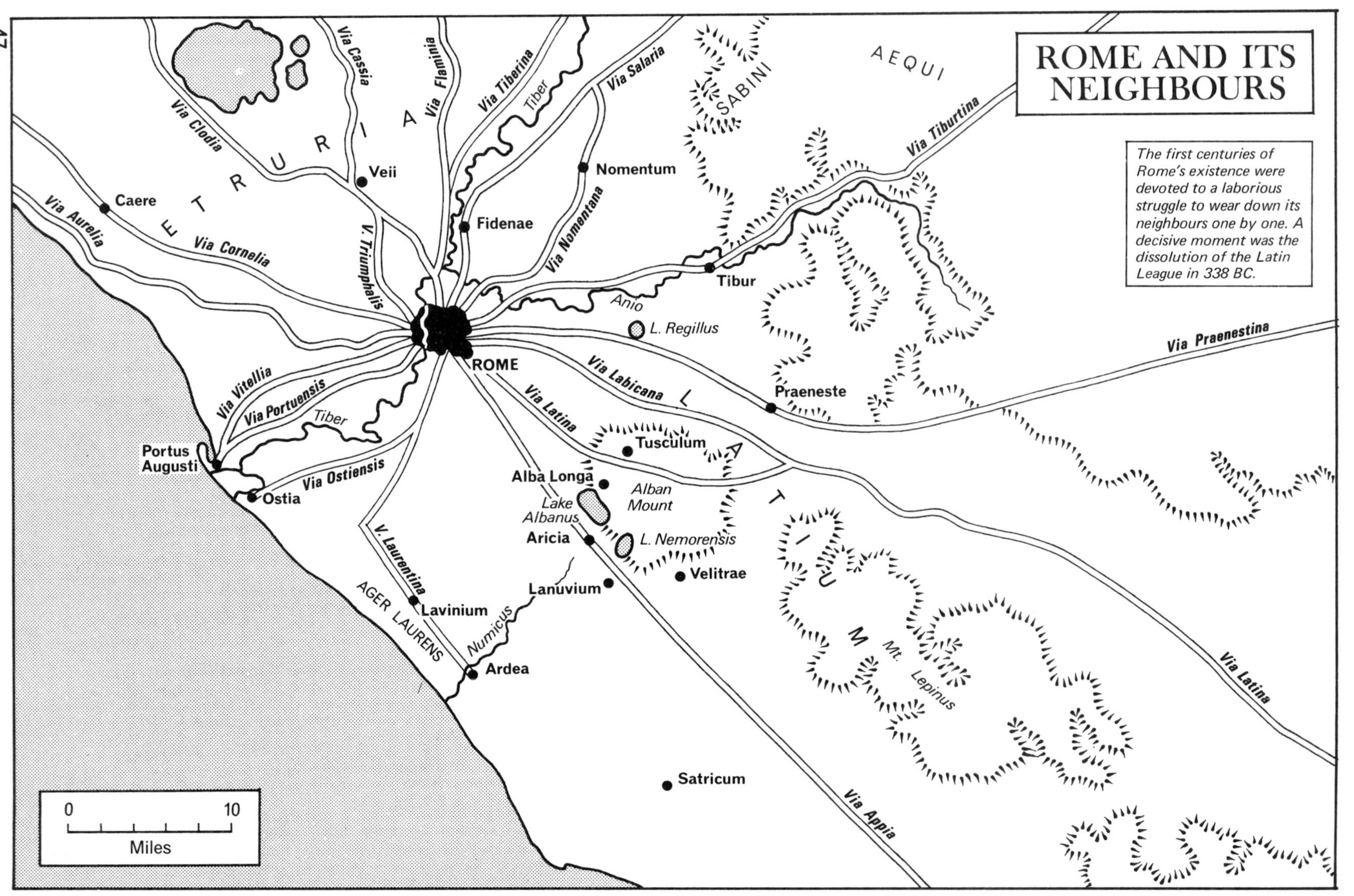
ROME AND ITS NEIGHBOURS
The first centuries of Rome's existence were devoted to a laborious struggle to wear down its neighbours one by one. A decisive moment was the dissolution of the Latin League in 338 BC.
AEQUI
SABINI
ETRURIA
LATIUM
Via Clodia
Via Cassia
Via Flaminia
Via Tiberina
Tiber
Via Salaria
Via Tiburtina
Veii
Nomentum
Caere
Via Aurelia
Via Cornelia
V. Triumphalis
Fidenae
Via Nomentana
Tibur
Anio
L. Regillus
Via Praenestina
ROME
Via Labicana
Praeneste
Via Vitellia
Via Portuensis
Tiber
Via Latina
Tusculum
Portus Augusti
Via Ostiensis
Ostia
Alba Longa
Alban Mount
Lake Albanus
Aricia
L. Nemorensis
V. Laurentina
Velitrae
Lanuvium
Lavinium
AGER LAURENS
Numicus
Ardea
Mt. Lepinus
Via Latina
Satricum
Via Appia
0
10
Miles

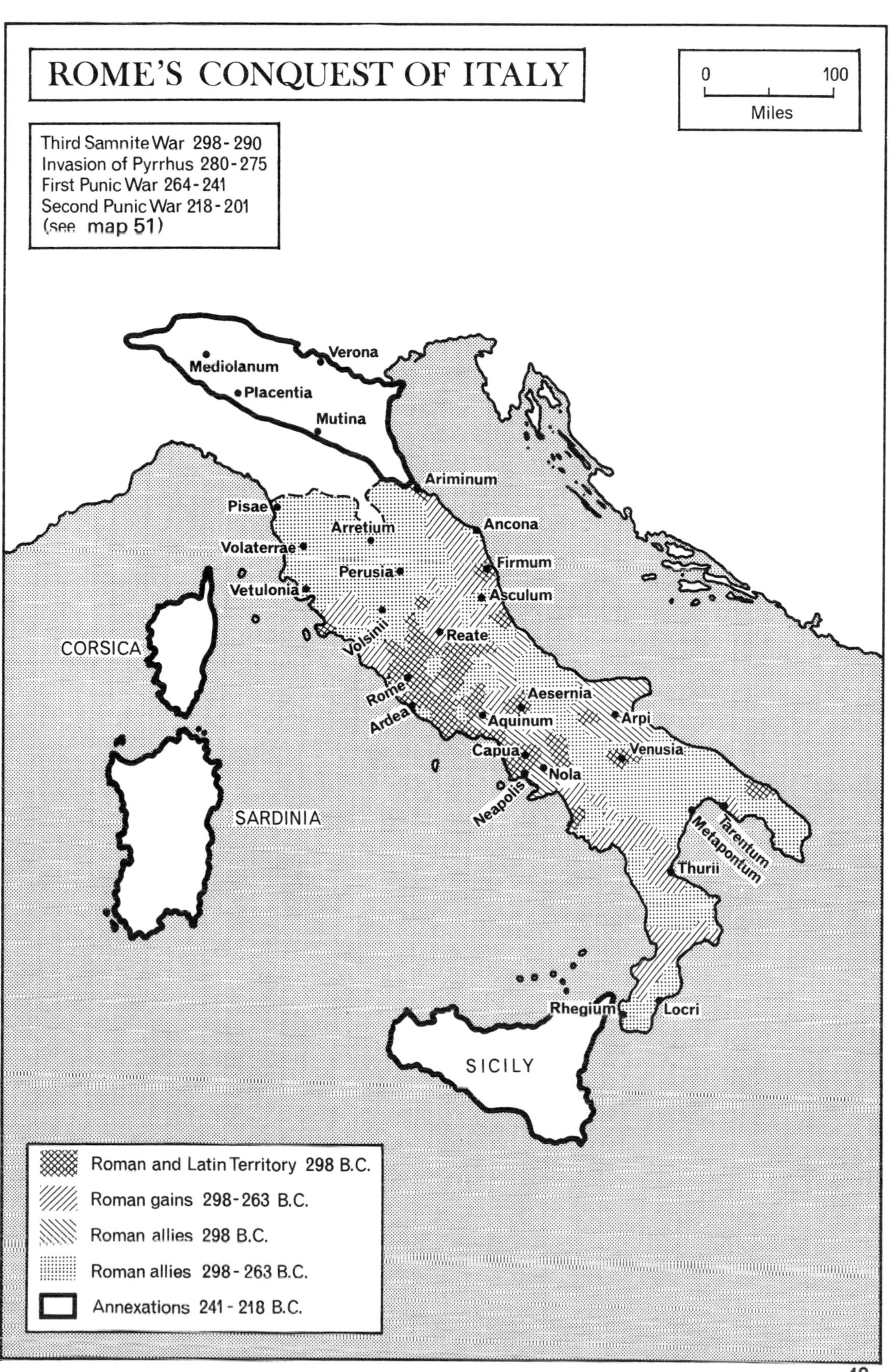
ROME'S CONQUEST OF ITALY
0
100
Miles
Third Samnite War 298-290
Invasion of Pyrrhus 280-275
First Punic War 264-241
Second Punic War 218-201
(see map 51)
Verona
Mediolanum
Placentia
Mutina
Ariminum
Pisae
Arretium
Ancona
Volaterrae
Firmum
Perusia
Vetulonia
Asculum
Volsinii
Reate
CORSICA
Rome
Aesernia
Ardea
Aquinum
Arpi
Capua
Venusia
Nola
Neapolis
SARDINIA
Tarentum
Metapontum
Thurii
Rhegium
Locri
SICILY
Roman and Latin Territory 298 B.C.
Roman gains 298-263 B.C.
Roman allies 298 B.C.
Roman allies 298-263 B.C.
Annexations 241-218 B.C.

THE ROADS OF ROMAN ITALY
0
100
Miles
Alps
Augusta Praetoria
Segusio
Mediolanum
Placentia
Cremona
Verona
Aquileia
Mantua
Dertona
Po
Genua
Ravenna
Ariminum
Fanum Fortunae
Luna
Florentia
Pisae
Apennines
Arretium
Vada Volaterrana
Tiber
Truentum
ADRIATIC SEA
Aternum
Reate
Tibur
CORSICA
Corfinium
ROME
Anagnia
Fregellae
Beneventum
Capua
Canusium
Tarracina
Cales
Casilinum
Neapolis
Venusia
Brundisium
Tarentum
SARDINIA
TYRRHENIAN SEA
Rhegium
SICILY
1 Via Aemilia (187 B.C.)
2 Via Appia (312 - 244 B.C.)
3 Via Aurelia
4 Via Flaminia (220 B.C.)
5 Via Latina
6 Via Postumia (148 B.C.)
7 Via Valeria
8 Via Julia Augusta
9 Via Domitiana
10 Via Trajana
11 Via Cassia
12 Via Popillia
13 Via Salaria

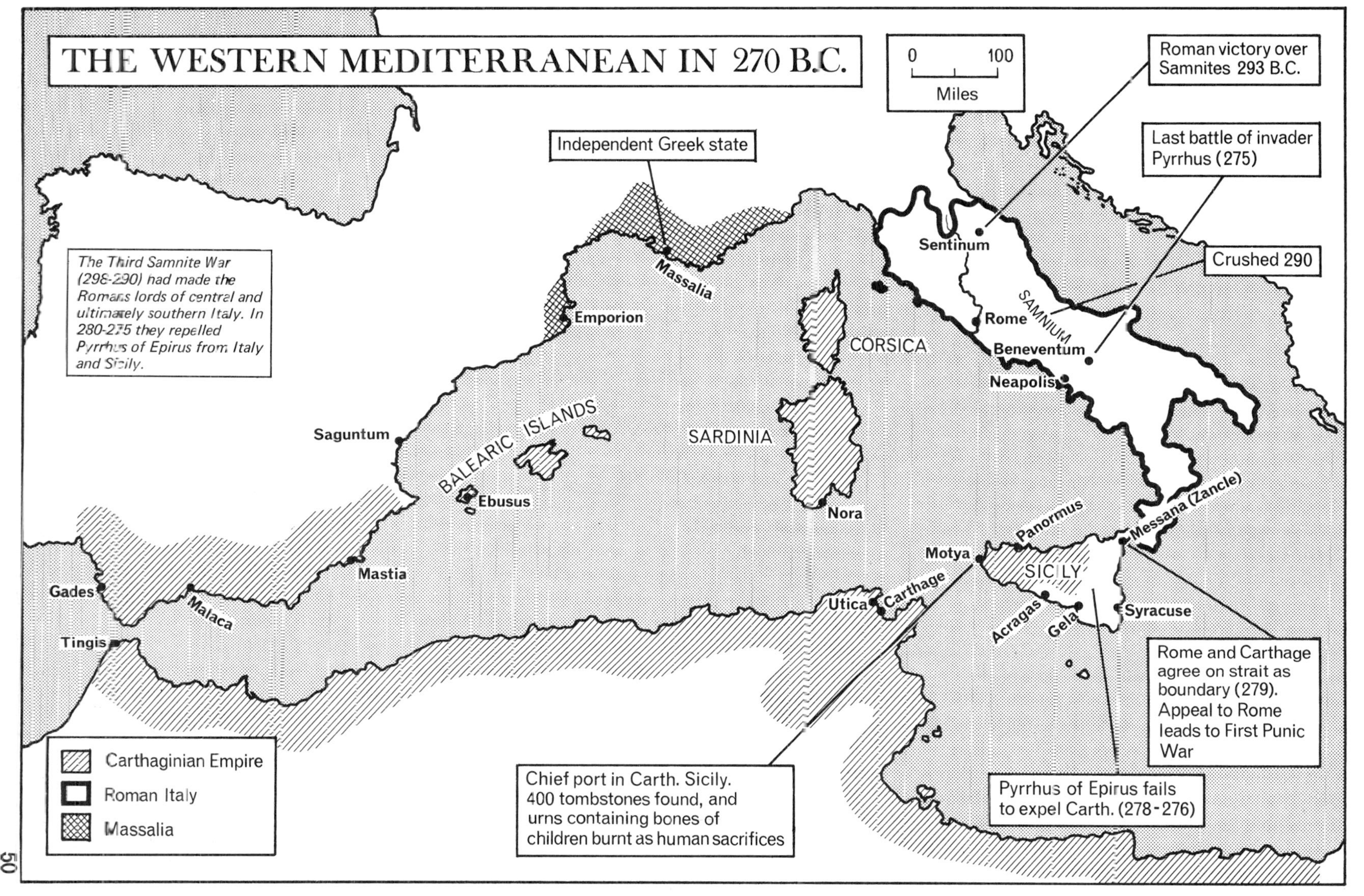
THE WESTERN MEDITERRANEAN IN 270 B.C.
0 100
Miles
Roman victory over Samnites 293 B.C.
Last battle of invader Pyrrhus (275)
Crushed 290
Independent Greek state
The Third Samnite War (298-290) had made the Romans lords of central and ultimately southern Italy. In 280-275 they repelled Pyrrhus of Epirus from Italy and Sicily.
Massalia
Emporion
Sentinum
Rome
SAMNIUM
Beneventum
Neapolis
CORSICA
SARDINIA
Nora
Saguntum
BALEARIC ISLANDS
Ebusus
Mastia
Gades
Malaca
Tingis
Motya
Panormus
Messana (Zancle)
SICILY
Acragas
Gela
Syracuse
Utica
Carthage
Rome and Carthage agree on strait as boundary (279). Appeal to Rome leads to First Punic War
Chief port in Carth. Sicily. 400 tombstones found, and urns containing bones of children burnt as human sacrifices
Pyrrhus of Epirus fails to expel Carth. (278-276)
Carthaginian Empire
Roman Italy
Massalia

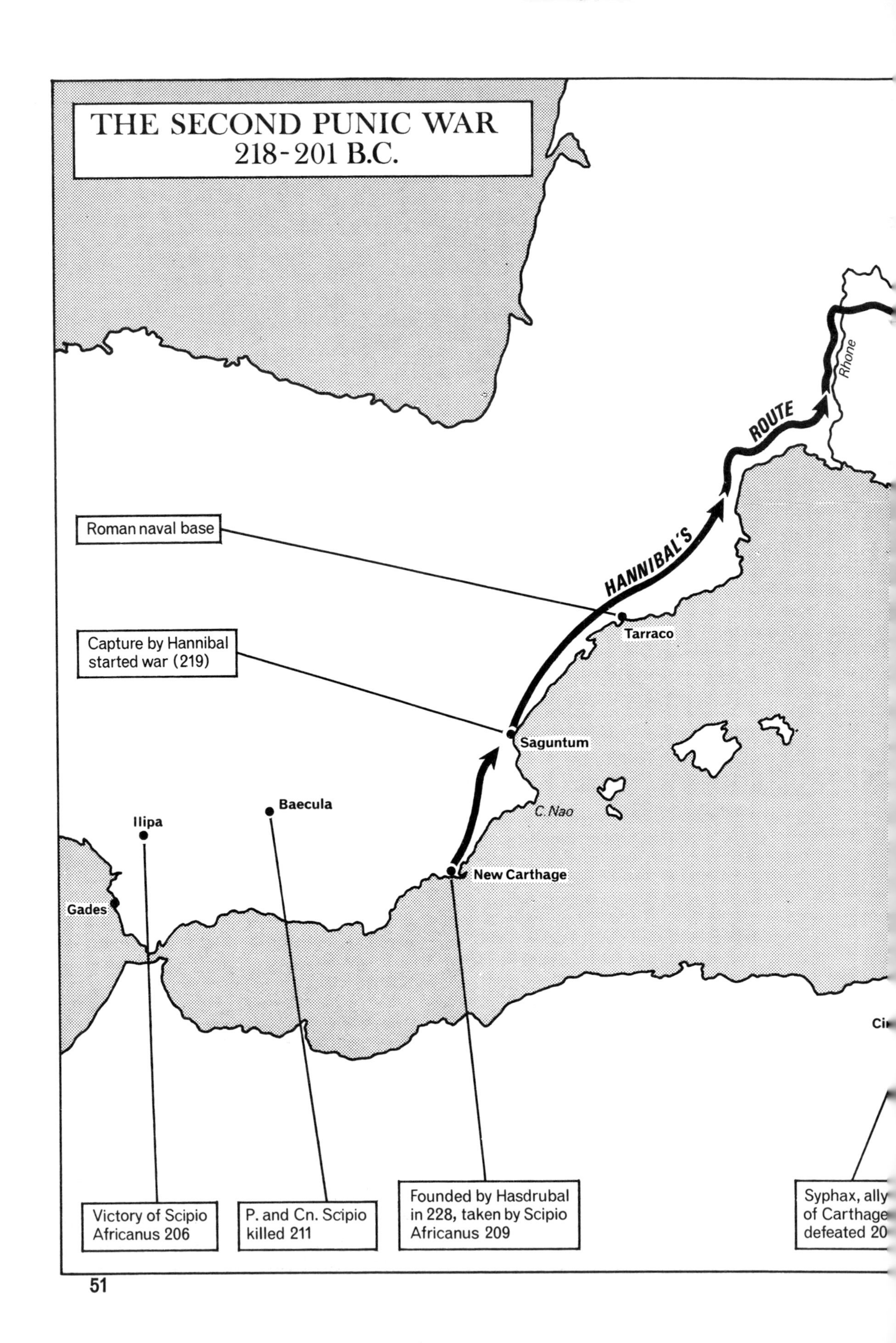
THE SECOND PUNIC WAR
218-201 B.C.
Rhone
HANNIBAL'S ROUTE
Roman naval base
Tarraco
Capture by Hannibal
started war (219)
Saguntum
C. Nao
Baecula
Ilipa
New Carthage
Gades
Victory of Scipio
Africanus 206
P. and Cn. Scipio
killed 211
Founded by Hasdrubal
in 228, taken by Scipio
Africanus 209
Syphax, ally
of Carthage
defeated 20

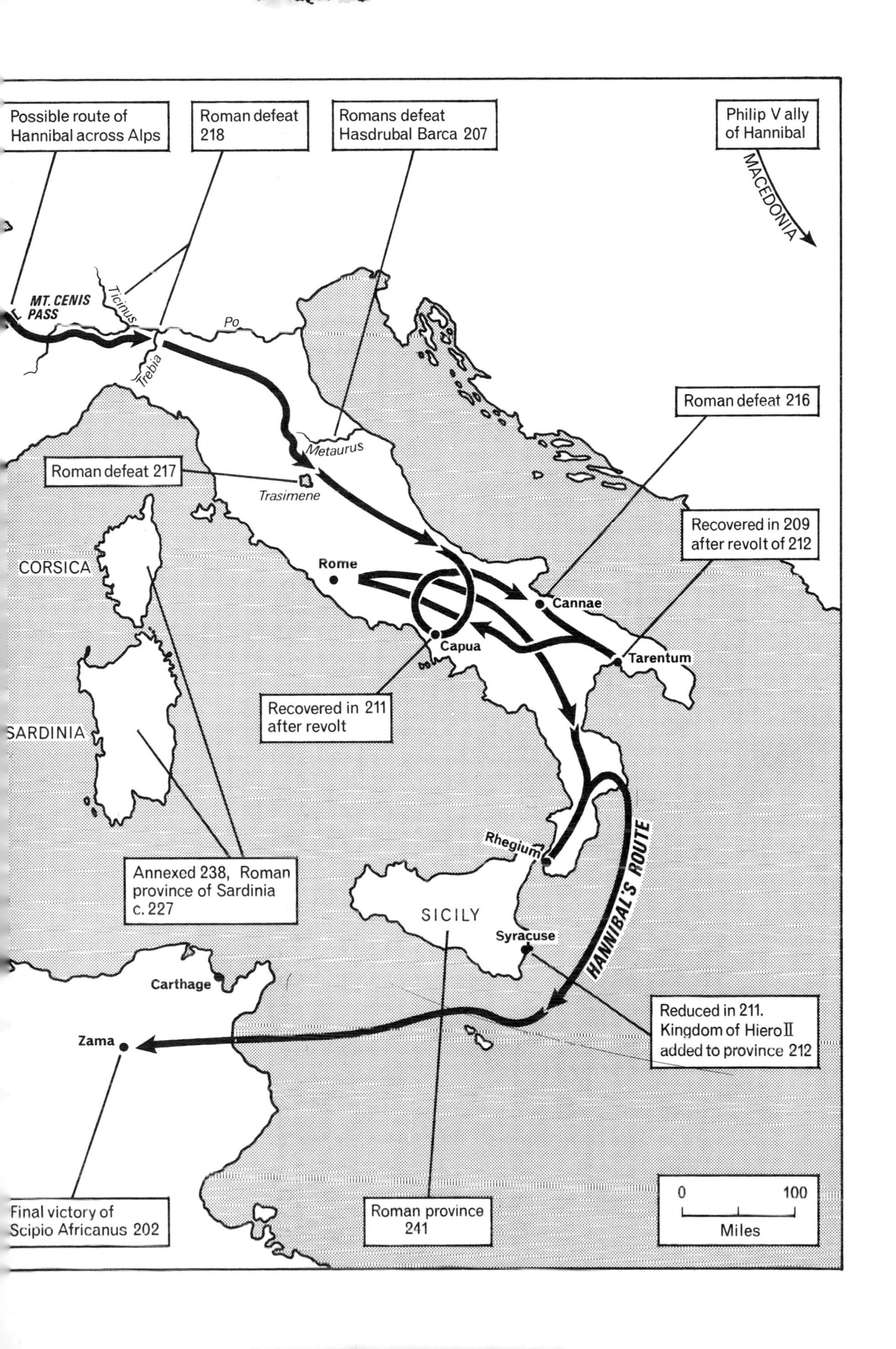

Possible route of Hannibal across Alps
Roman defeat 218
Romans defeat Hasdrubal Barca 207
Philip V ally of Hannibal
MACEDONIA
MT. CENIS PASS
Ticinus
Po
Trebia
Metaurus
Roman defeat 216
Roman defeat 217
Trasimene
Recovered in 209 after revolt of 212
CORSICA
Rome
Cannae
Capua
Tarentum
Recovered in 211 after revolt
SARDINIA
Rhegium
HANNIBAL'S ROUTE
Annexed 238, Roman province of Sardinia c. 227
SICILY
Syracuse
Carthage
Reduced in 211. Kingdom of Hiero II added to province 212
Zama
Final victory of Scipio Africanus 202
Roman province 241
0
100
Miles

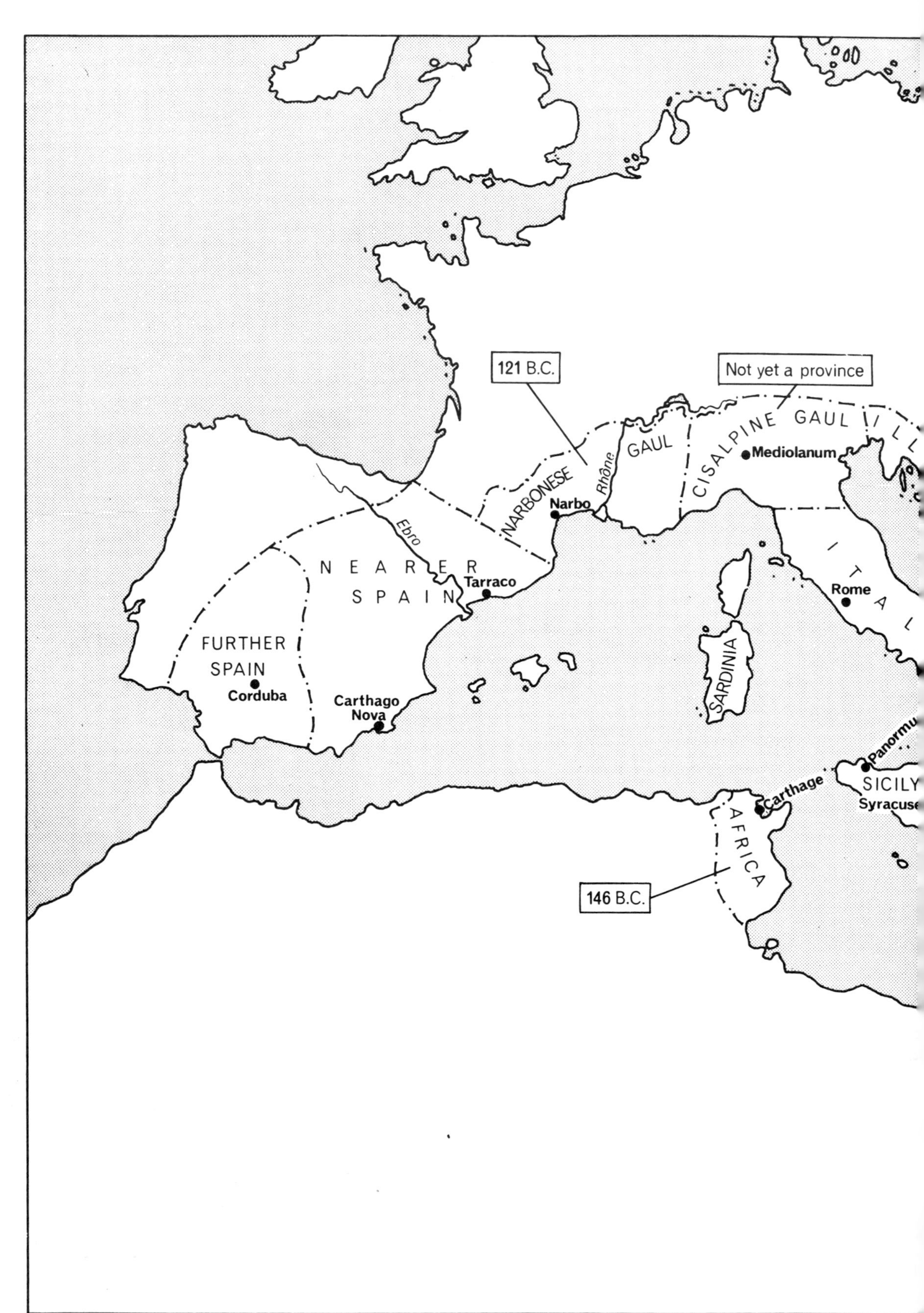
121 B.C.
Not yet a province
NARBONESE
GAUL
Rhône
CISALPINE GAUL
Mediolanum
Narbo
Ebro
NEARER SPAIN
Tarraco
FURTHER SPAIN
Corduba
Carthago Nova
SARDINIA
Rome
ITAL
SICILY
Syracuse
Carthage
AFRICA
146 B.C.

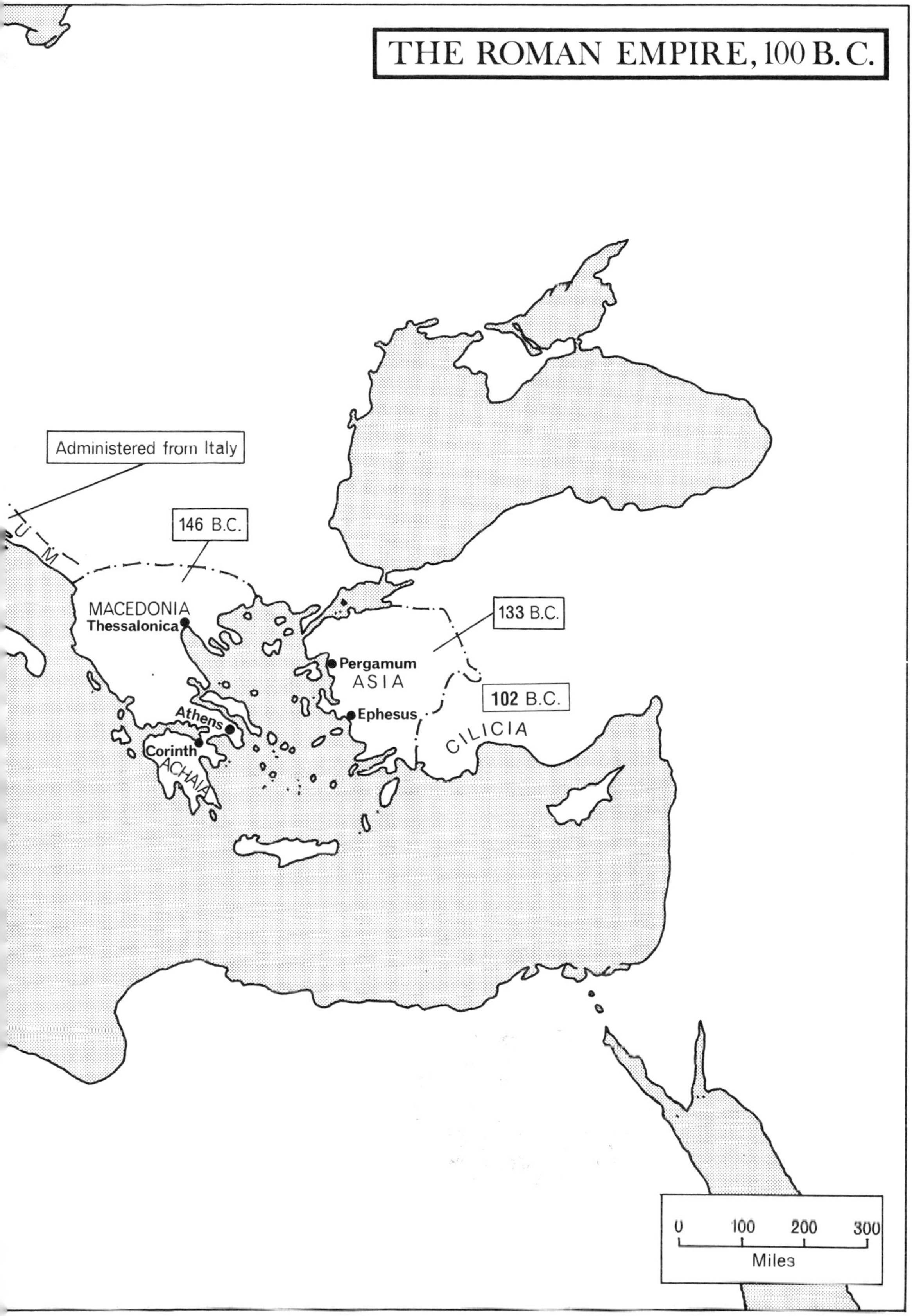

THE ROMAN EMPIRE, 100 B.C.

Administered from Italy

146 B.C.

MACEDONIA

Thessalonica

133 B.C.

Pergamum

ASIA

Ephesus

102 B.C.

CILICIA

Athens

Corinth

ACHAIA

0 100 200 300

Miles

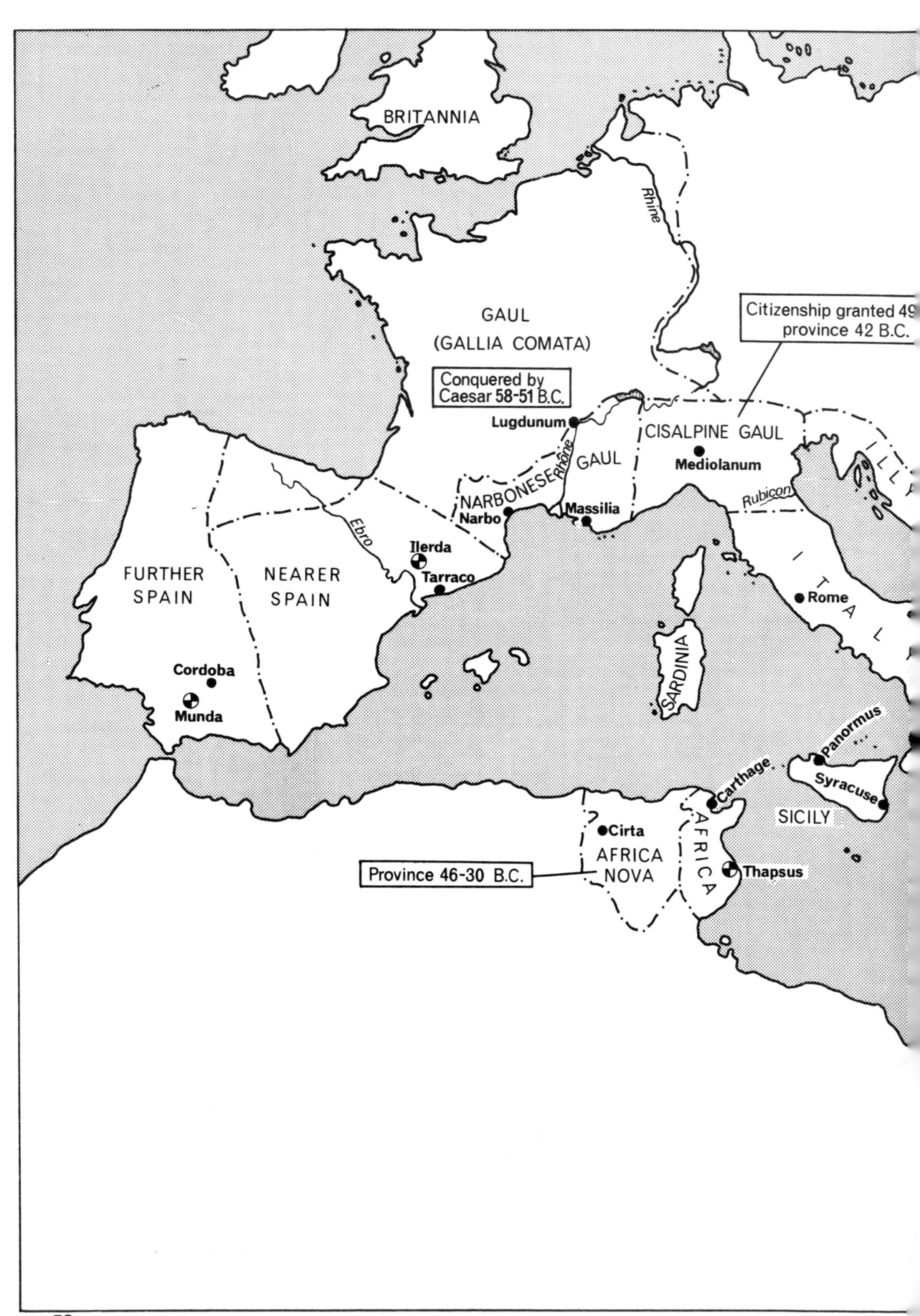
BRITANNIA
Rhine
GAUL
(GALLIA COMATA)
Conquered by
Caesar 58-51 B.C.
Citizenship granted 49
province 42 B.C.
Lugdunum
CISALPINE GAUL
Mediolanum
GAUL
Rhone
NARBONESE
Narbo
Massilia
Rubicon
Ebro
Ilerda
Tarraco
FURTHER
SPAIN
NEARER
SPAIN
Rome
I T A L
SARDINIA
Cordoba
Munda
Panormus
Carthage
Syracuse
SICILY
Cirta
AFRICA
NOVA
AFRICA
Province 46-30 B.C.
Thapsus

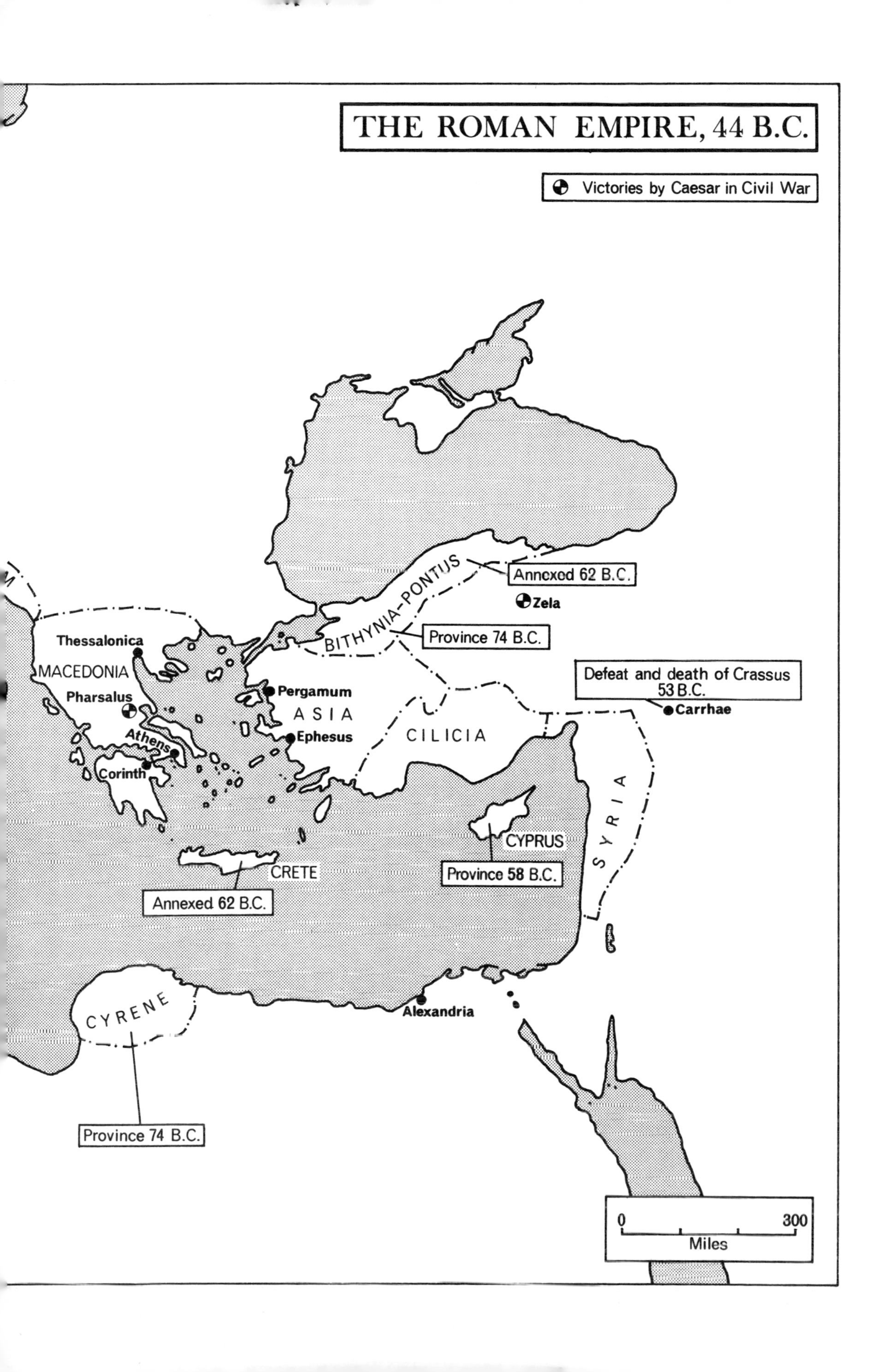

THE ROMAN EMPIRE, 44 B.C.
Victories by Caesar in Civil War
BITHYNIA-PONTUS
Annexed 62 B.C.
Zela
Province 74 B.C.
Thessalonica
MACEDONIA
Pharsalus
Pergamum
ASIA
Athens
Ephesus
Corinth
CILICIA
Defeat and death of Crassus 53 B.C.
Carrhae
SYRIA
CYPRUS
Province 58 B.C.
CRETE
Annexed 62 B.C.
CYRENE
Alexandria
Province 74 B.C.
0
300
Miles

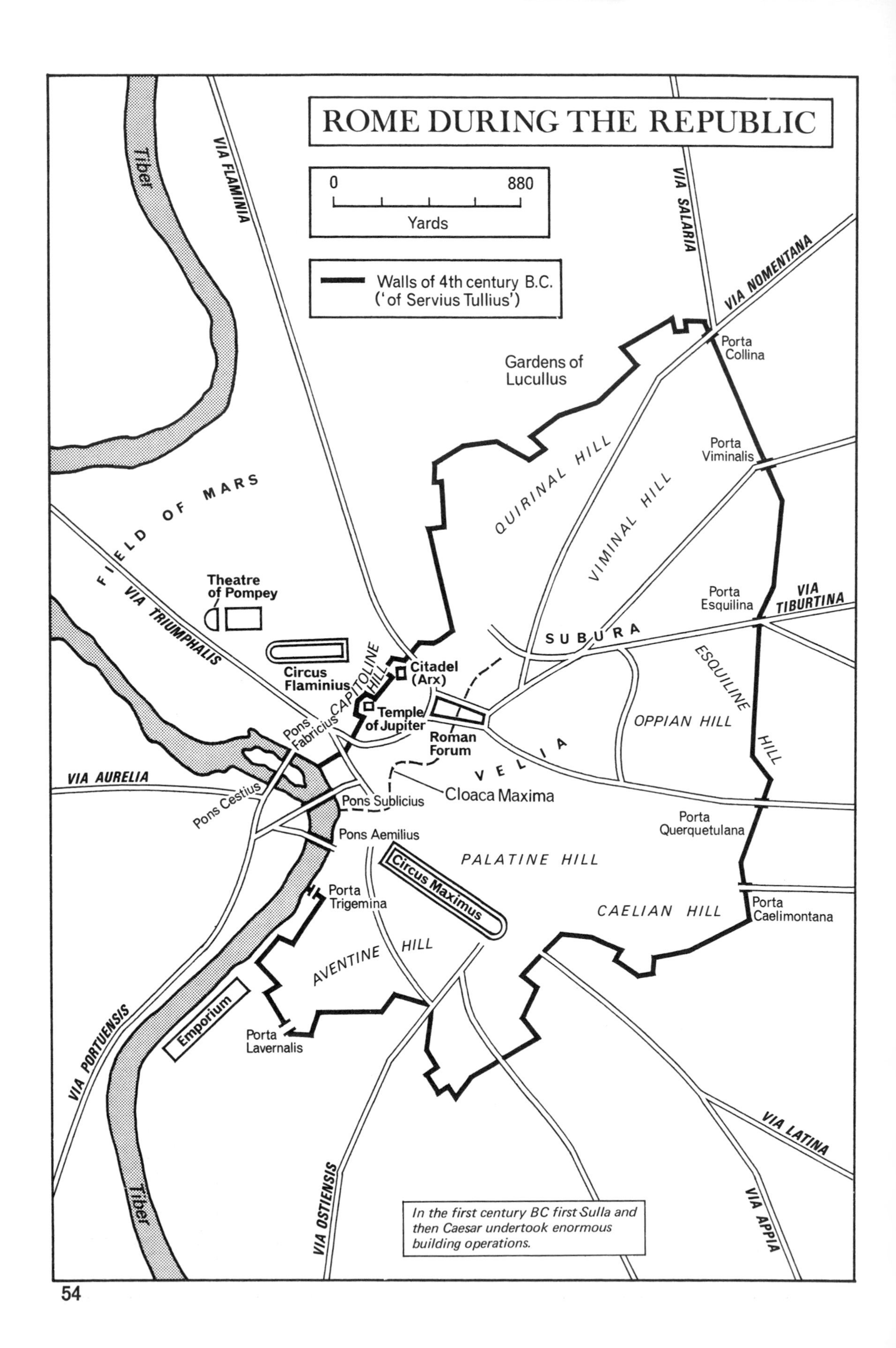

ROME DURING THE REPUBLIC
0
880
Yards
Walls of 4th century B.C.
('of Servius Tullius')
Tiber
VIA FLAMINIA
VIA SALARIA
VIA NOMENTANA
Porta Collina
Gardens of Lucullus
Porta Viminalis
QUIRINAL HILL
VIMINAL HILL
FIELD OF MARS
Theatre of Pompey
VIA TRIUMPHALIS
Porta Esquilina
VIA TIBURTINA
SUBURA
Circus Flaminius
CAPITOLINE HILL
Citadel (Arx)
Temple of Jupiter
Roman Forum
ESQUILINE HILL
OPPIAN HILL
Pons Fabricius
VELIA
VIA AURELIA
Pons Cestius
Pons Sublicius
Cloaca Maxima
Pons Aemilius
Porta Querquetulana
PALATINE HILL
Circus Maximus
Porta Trigemina
CAELIAN HILL
Porta Caelimontana
AVENTINE HILL
Emporium
Porta Lavernalis
VIA PORTUENSIS
VIA LATINA
VIA OSTIENSIS
VIA APPIA
Tiber
In the first century BC first Sulla and then Caesar undertook enormous building operations.

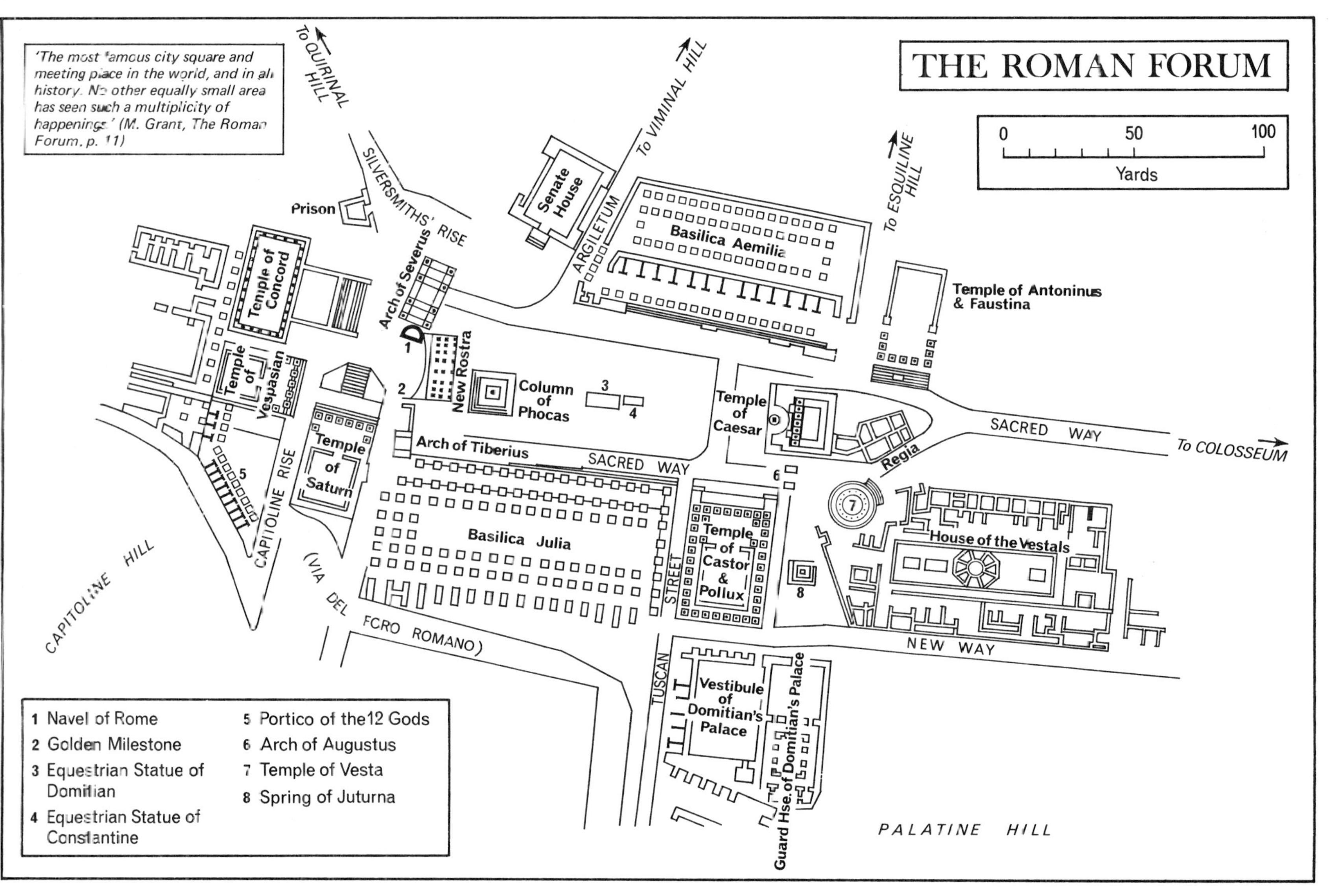

THE ROMAN FORUM
0
50
100
Yards
'The most famous city square and meeting place in the world, and in all history. No other equally small area has seen such a multiplicity of happenings.' (M. Grant, The Roman Forum, p. 11)
To QUIRINAL HILL
To VIMINAL HILL
To ESQUILINE HILL
To COLOSSEUM
Prison
SILVERSMITHS' RISE
Senate House
ARGILETUM
Basilica Aemilia
Temple of Antoninus & Faustina
Temple of Concord
Arch of Severus
New Rostra
Column of Phocas
Temple of Vespasian
Temple of Saturn
Temple of Caesar
Regia
SACRED WAY
Arch of Tiberius
SACRED WAY
CAPITOLINE RISE
CAPITOLINE HILL
Basilica Julia
STREET
TUSCAN
Temple of Castor & Pollux
House of the Vestals
NEW WAY
(VIA DEL FORO ROMANO)
Vestibule of Domitian's Palace
Guard Hse. of Domitian's Palace
PALATINE HILL
1 Navel of Rome
2 Golden Milestone
3 Equestrian Statue of Domitian
4 Equestrian Statue of Constantine
5 Portico of the 12 Gods
6 Arch of Augustus
7 Temple of Vesta
8 Spring of Juturna

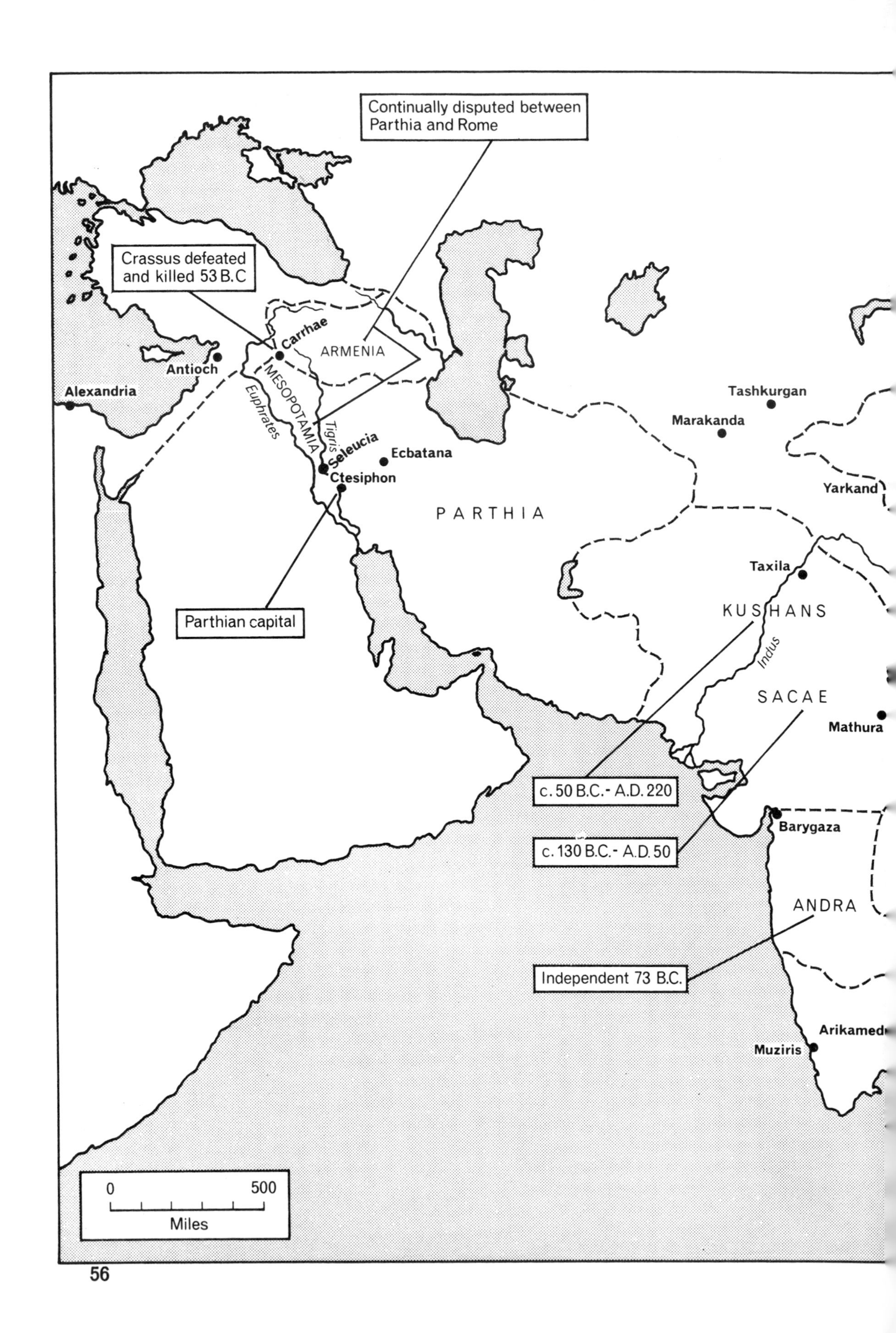
Continually disputed between Parthia and Rome
Crassus defeated and killed 53 B.C
Carrhae
ARMENIA
Antioch
Alexandria
MESOPOTAMIA
Euphrates
Tigris
Seleucia
Ctesiphon
Ecbatana
PARTHIA
Tashkurgan
Marakanda
Yarkand
Taxila
KUSHANS
Indus
SACAE
Mathura
Parthian capital
c. 50 B.C.- A.D. 220
c. 130 B.C.- A.D. 50
Barygaza
ANDRA
Independent 73 B.C.
Muziris
0
500
Miles

PARTHIA AND THE EAST

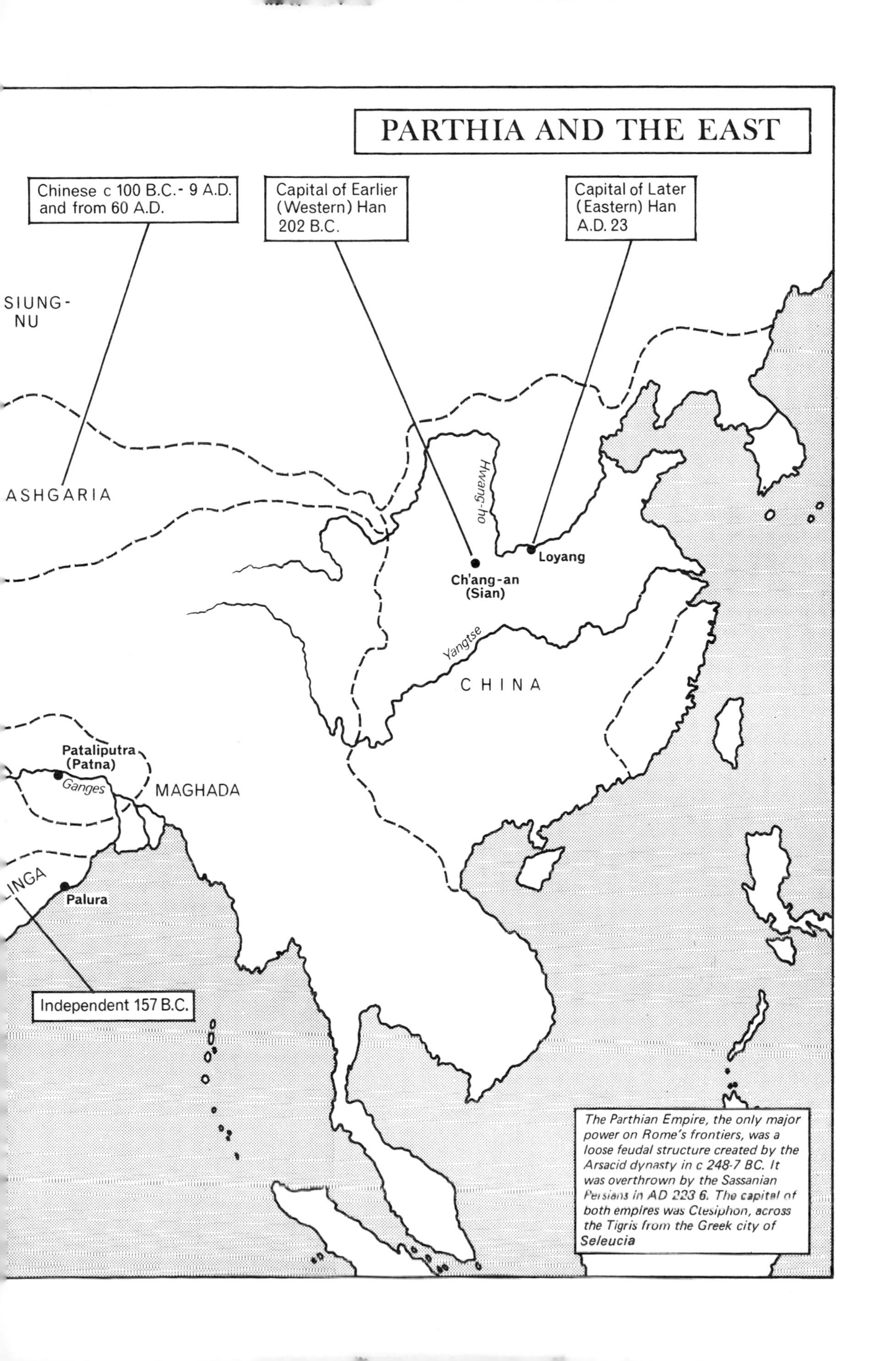

The Parthian Empire, the only major power on Rome's frontiers, was a loose feudal structure created by the Arsacid dynasty in c 248-7 BC. It was overthrown by the Sassanian Persians in AD 223 6. The capital of both empires was Ctesiphon, across the Tigris from the Greek city of Seleucia

BRITANNIA
FREE GERMANY
LWR. GERMANY (17 B.C.)
Temporarily conquered from 15 B.C. but abandoned after ambushing of Varus by Arminius in A.D. 9
Rhine
Colonia Agrippinensis
Moguntiacum
BELGICA
Danube
LOWER PANNONIA (10 B.C.
LUGDUNENSIS
UPR. GERMANY (17 B.C.)
RHAETIA (15 B.C.)
NORICUM (15 B.C.)
UPPER PANNONIA
P
Aquileia
Lugdunum
AQUITANIA
NARBONENSIS
C
M
Nemausus
ITALY
Adriatic Sea
Rome
TARRACONENSIS
Tarraco
LUSITANIA (c. 27 B.C.)
Corduba
BAETICA
Gades
Naulochus
SICILY
Carthage
MAURETANIA
Naval victory over Sextus Pompeius 36 B.C.
AFRICA
Imperial frontier as in A.D. 14
Provincial frontiers
ASIA Senatorial provinces
ALPINE PROVINCES (15-14 B.C.)
M: Maritime, C: Cottian, P: Pennine
The hatched areas represent the more important dependent ('client') states, whose monarchs enjoyed internal autonomy but had to support Rome's foreign policy and help defend the imperial frontiers.
Principal client states

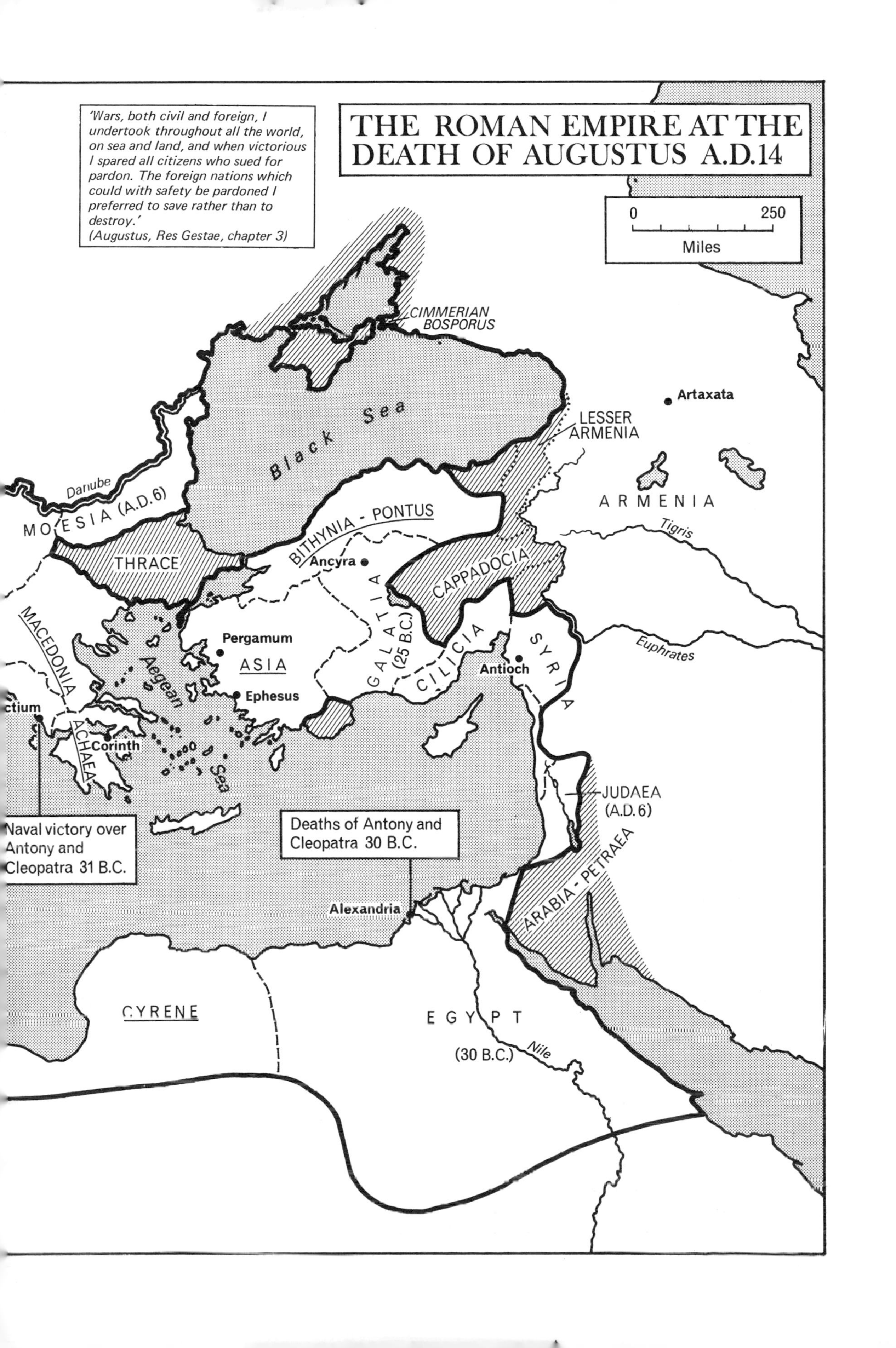
THE ROMAN EMPIRE AT THE DEATH OF AUGUSTUS A.D.14
'Wars, both civil and foreign, I undertook throughout all the world, on sea and land, and when victorious I spared all citizens who sued for pardon. The foreign nations which could with safety be pardoned I preferred to save rather than to destroy.' (Augustus, Res Gestae, chapter 3)
0
250
Miles
CIMMERIAN BOSPORUS
Black Sea
Artaxata
LESSER ARMENIA
ARMENIA
Tigris
Euphrates
Danube
MOESIA (A.D.6)
THRACE
BITHYNIA - PONTUS
Ancyra
CAPPADOCIA
GALATIA (25 B.C.)
MACEDONIA
Pergamum
ASIA
Ephesus
Aegean Sea
CILICIA
Antioch
SYRIA
ctium
ACHAEA
Corinth
JUDAEA (A.D.6)
Naval victory over Antony and Cleopatra 31 B.C.
Deaths of Antony and Cleopatra 30 B.C.
ARABIA - PETRAEA
Alexandria
CYRENE
EGYPT (30 B.C.)
Nile

All roads lead to Rome: the most potent guarantees of external and internal peace and stimulants of prosperity.

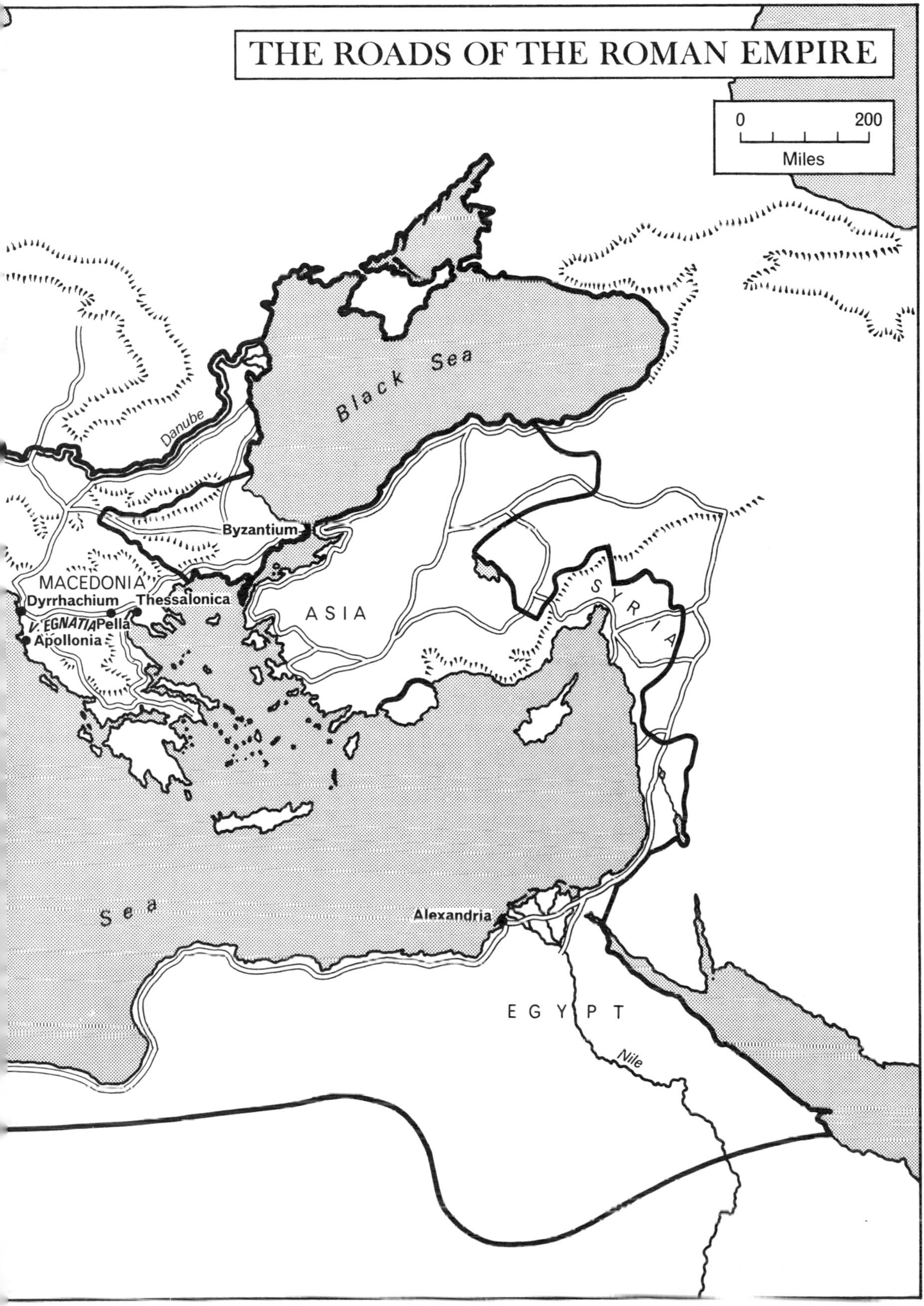

THE ROADS OF THE ROMAN EMPIRE

0 200
Miles

Black Sea
Danube
Byzantium
MACEDONIA
Dyrrhachium
Thessalonica
V. EGNATIA
Pella
Apollonia
ASIA
SYRIA
Sea
Alexandria
EGYPT
Nile

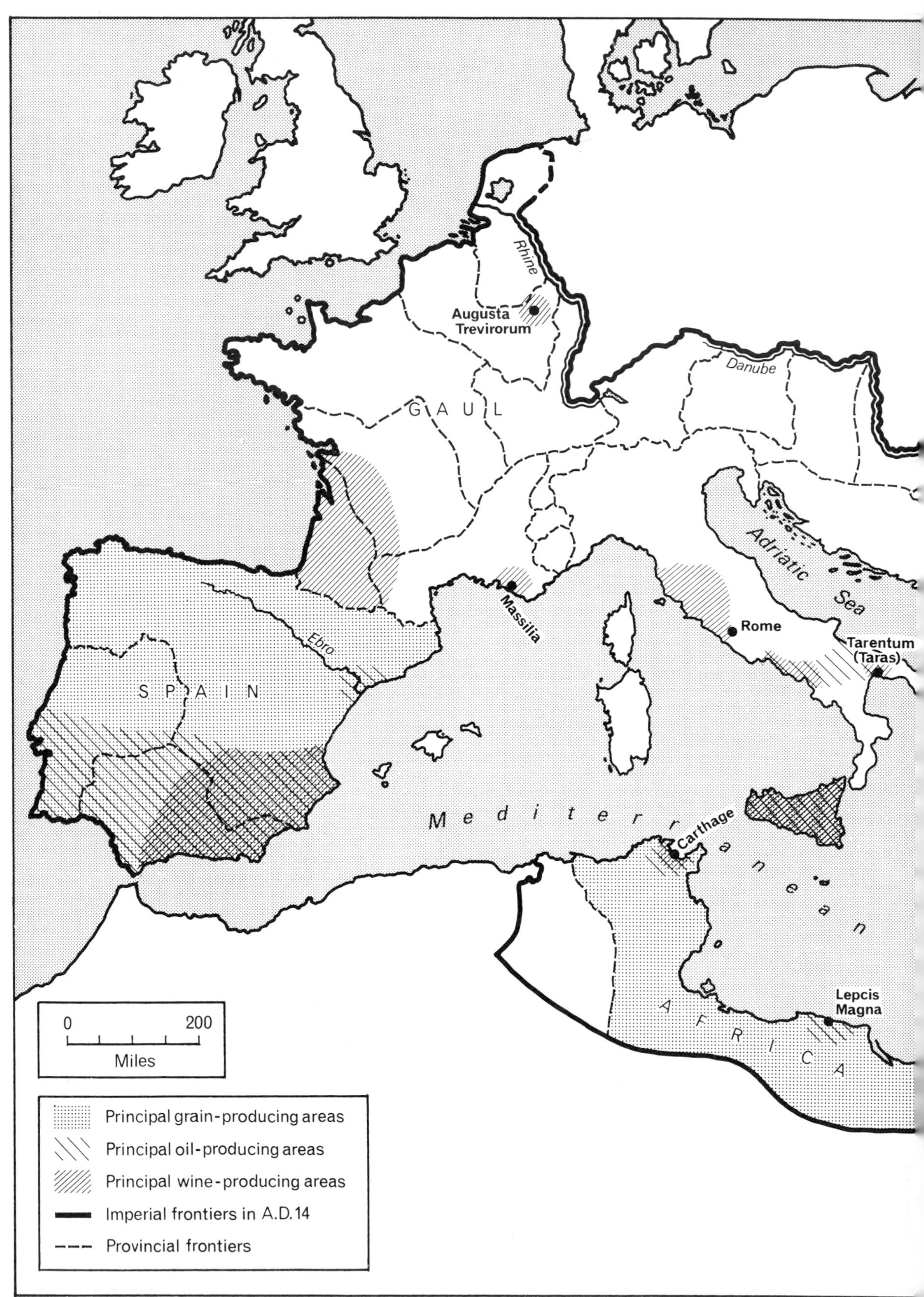

Rhine
Augusta
Trevirorum
Danube
G A U L
Adriatic
Sea
Massilia
Rome
Tarentum
(Taras)
Ebro
S P A I N
M e d i t e r r a n e a n
Carthage
Lepcis
Magna
A F R I C A
0
200
Miles
Principal grain-producing areas
Principal oil-producing areas
Principal wine-producing areas
Imperial frontiers in A.D.14
Provincial frontiers

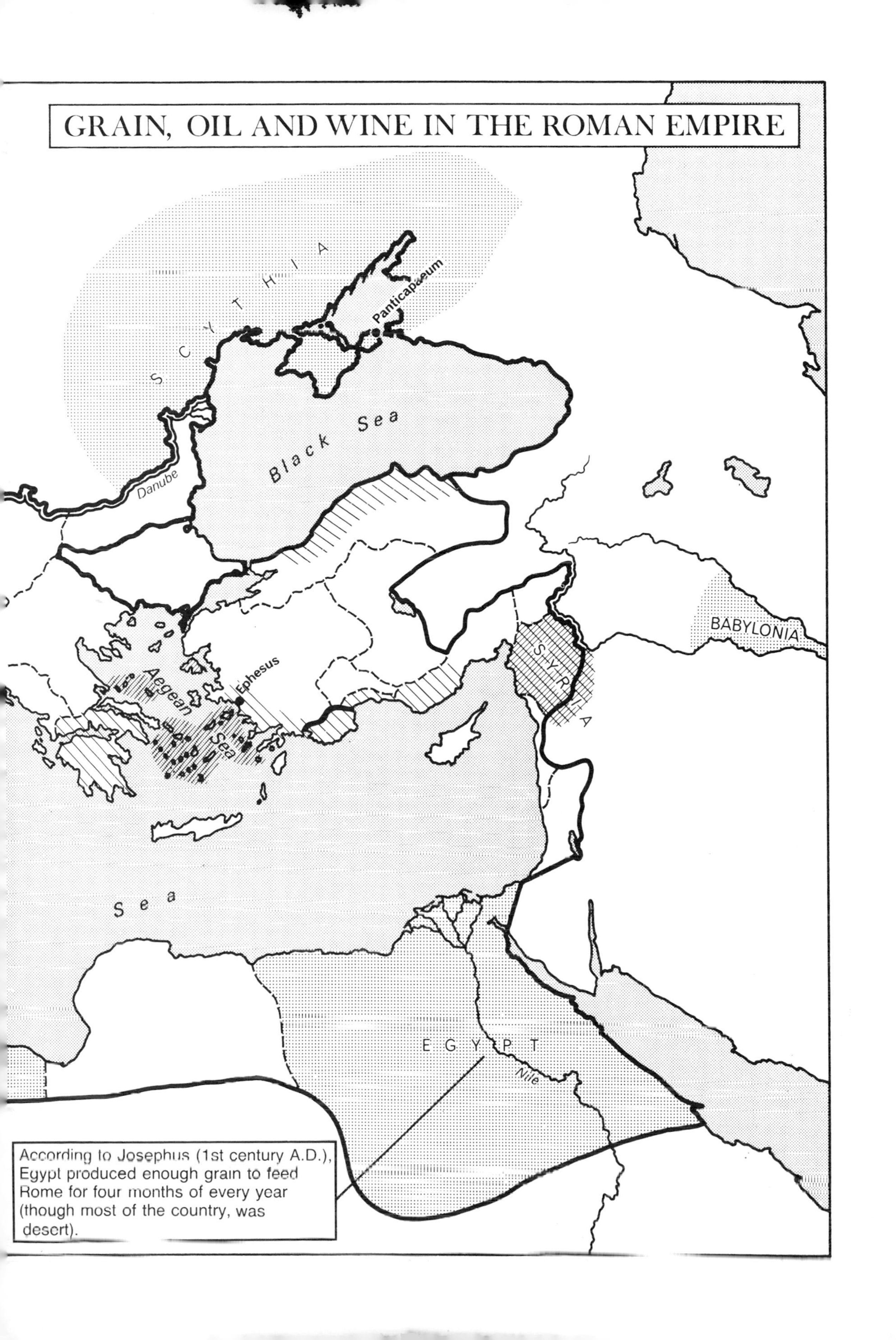
GRAIN, OIL AND WINE IN THE ROMAN EMPIRE
S C Y T H I A
Panticapaeum
Black Sea
Danube
Aegean Sea
Ephesus
SYRIA
BABYLONIA
Sea
EGYPT
Nile
According to Josephus (1st century A.D.), Egypt produced enough grain to feed Rome for four months of every year (though most of the country, was desert).

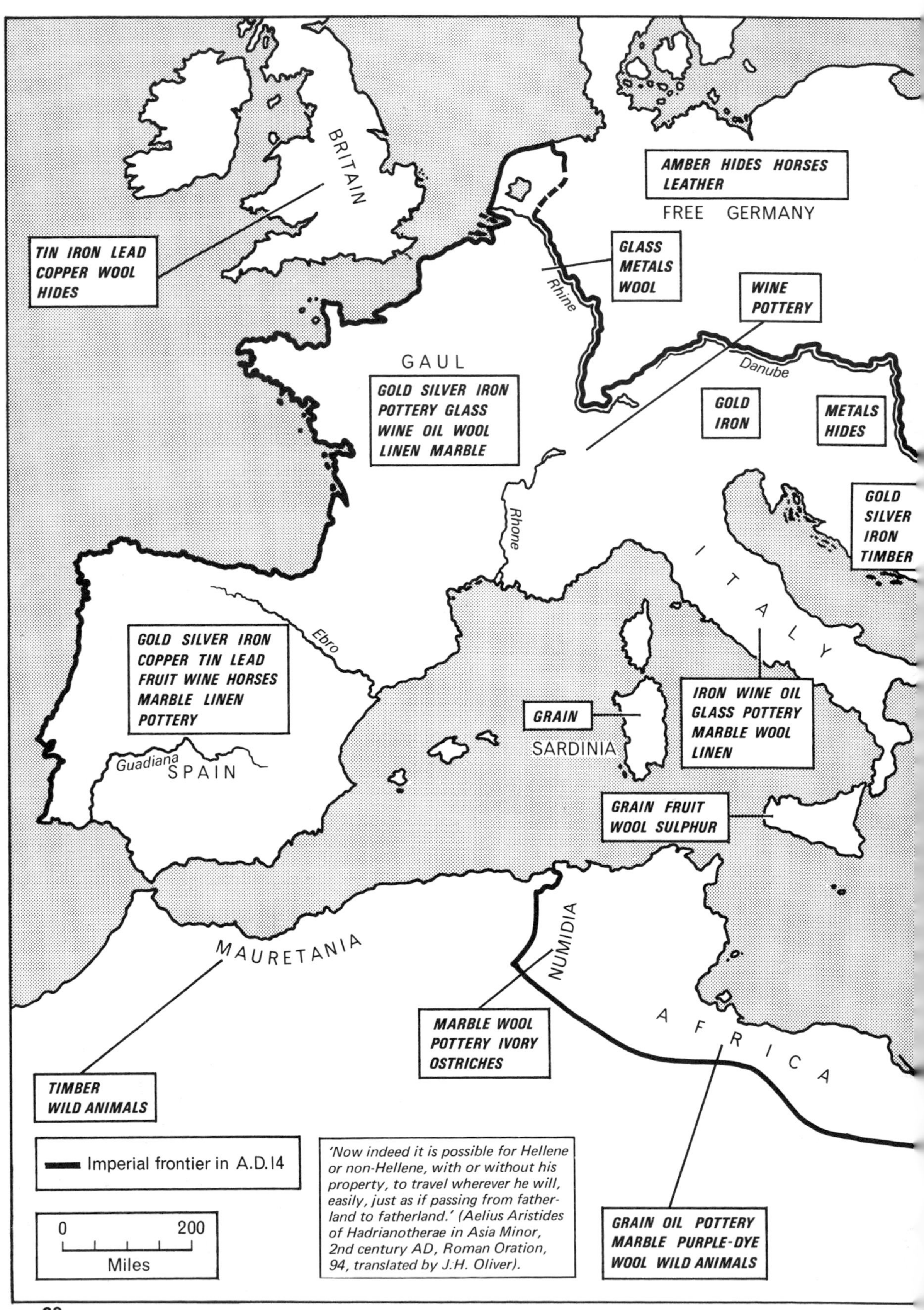

BRITAIN
TIN IRON LEAD
COPPER WOOL
HIDES
AMBER HIDES HORSES
LEATHER
FREE GERMANY
GLASS
METALS
WOOL
Rhine
WINE
POTTERY
GAUL
GOLD SILVER IRON
POTTERY GLASS
WINE OIL WOOL
LINEN MARBLE
Danube
GOLD
IRON
METALS
HIDES
GOLD
SILVER
IRON
TIMBER
Rhone
ITALY
Ebro
GOLD SILVER IRON
COPPER TIN LEAD
FRUIT WINE HORSES
MARBLE LINEN
POTTERY
GRAIN
SARDINIA
IRON WINE OIL
GLASS POTTERY
MARBLE WOOL
LINEN
Guadiana
SPAIN
GRAIN FRUIT
WOOL SULPHUR
NUMIDIA
MAURETANIA
MARBLE WOOL
POTTERY IVORY
OSTRICHES
AFRICA
TIMBER
WILD ANIMALS
Imperial frontier in A.D.14
0
200
Miles
'Now indeed it is possible for Hellene or non-Hellene, with or without his property, to travel wherever he will, easily, just as if passing from fatherland to fatherland.' (Aelius Aristides of Hadrianotherae in Asia Minor, 2nd century AD, Roman Oration, 94, translated by J.H. Oliver).
GRAIN OIL POTTERY
MARBLE PURPLE-DYE
WOOL WILD ANIMALS

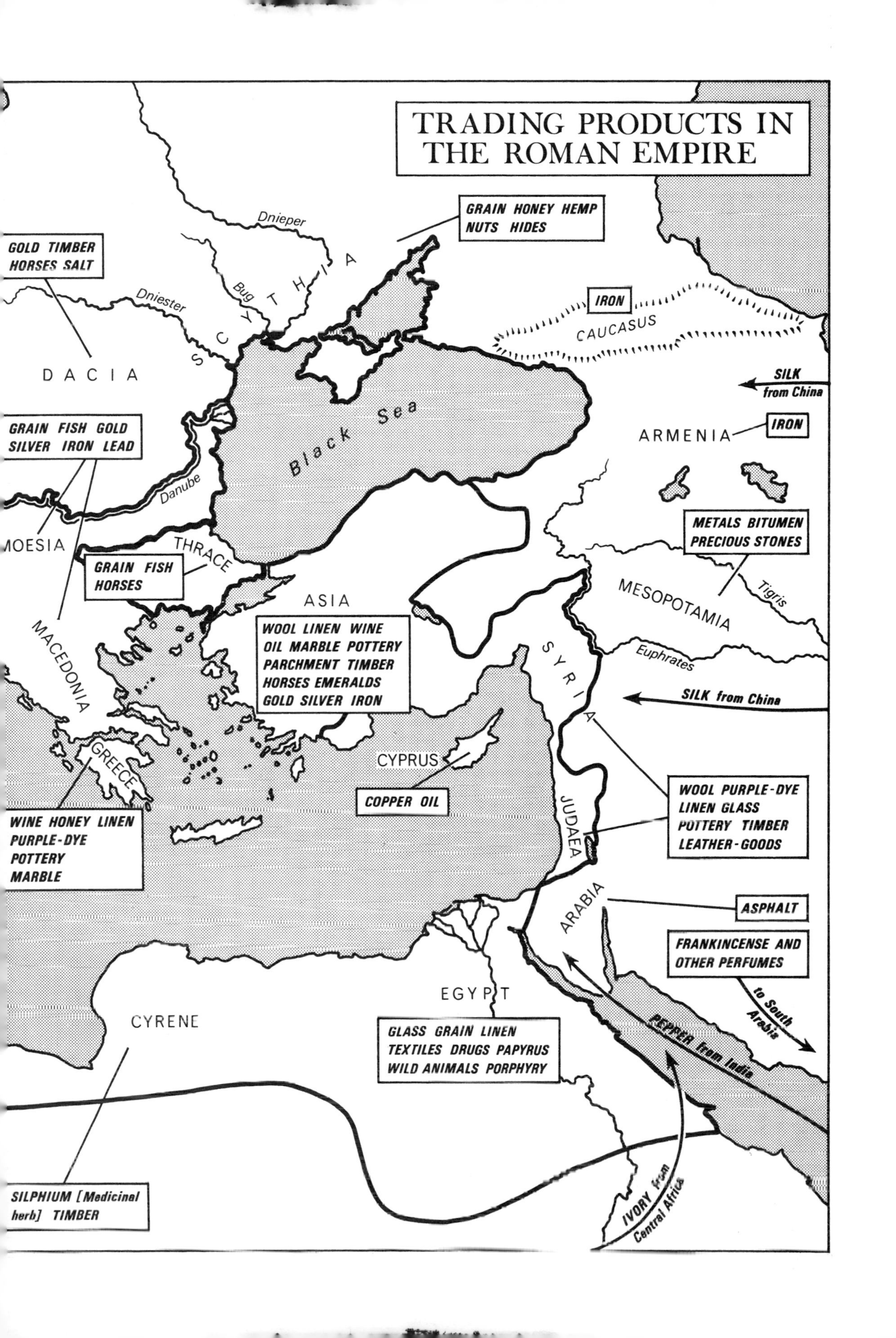
TRADING PRODUCTS IN THE ROMAN EMPIRE
GRAIN HONEY HEMP NUTS HIDES
GOLD TIMBER HORSES SALT
Dnieper
Dniester
Bug
SCYTHIA
IRON
CAUCASUS
DACIA
SILK from China
GRAIN FISH GOLD SILVER IRON LEAD
Black Sea
ARMENIA
IRON
Danube
MOESIA
THRACE
METALS BITUMEN PRECIOUS STONES
GRAIN FISH HORSES
ASIA
MESOPOTAMIA
Tigris
MACEDONIA
WOOL LINEN WINE OIL MARBLE POTTERY PARCHMENT TIMBER HORSES EMERALDS GOLD SILVER IRON
SYRIA
Euphrates
SILK from China
GREECE
CYPRUS
COPPER OIL
WOOL PURPLE-DYE LINEN GLASS POTTERY TIMBER LEATHER-GOODS
JUDAEA
WINE HONEY LINEN PURPLE-DYE POTTERY MARBLE
ARABIA
ASPHALT
FRANKINCENSE AND OTHER PERFUMES
to South Arabia
EGYPT
PEPPER from India
CYRENE
GLASS GRAIN LINEN TEXTILES DRUGS PAPYRUS WILD ANIMALS PORPHYRY
SILPHIUM [Medicinal herb] TIMBER
IVORY from Central Africa

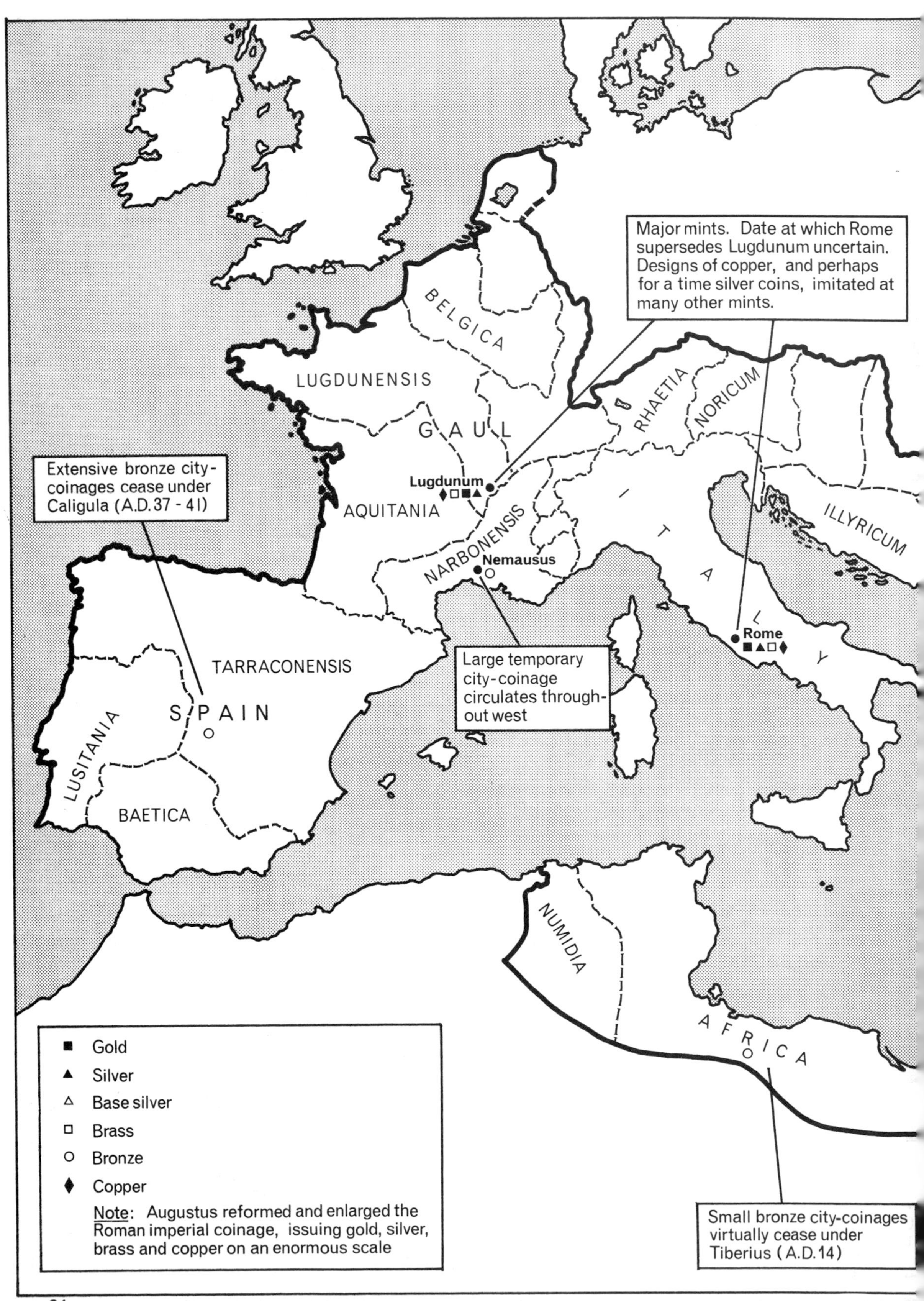
Major mints. Date at which Rome supersedes Lugdunum uncertain. Designs of copper, and perhaps for a time silver coins, imitated at many other mints.
BELGICA
LUGDUNENSIS
GAUL
RHAETIA
NORICUM
Extensive bronze city-coinages cease under Caligula (A.D. 37 - 41)
Lugdunum
AQUITANIA
NARBONENSIS
ITALY
ILLYRICUM
Nemausus
Rome
Large temporary city-coinage circulates throughout west
TARRACONENSIS
SPAIN
LUSITANIA
BAETICA
NUMIDIA
AFRICA
Gold
Silver
Base silver
Brass
Bronze
Copper
Note: Augustus reformed and enlarged the Roman imperial coinage, issuing gold, silver, brass and copper on an enormous scale
Small bronze city-coinages virtually cease under Tiberius (A.D. 14)

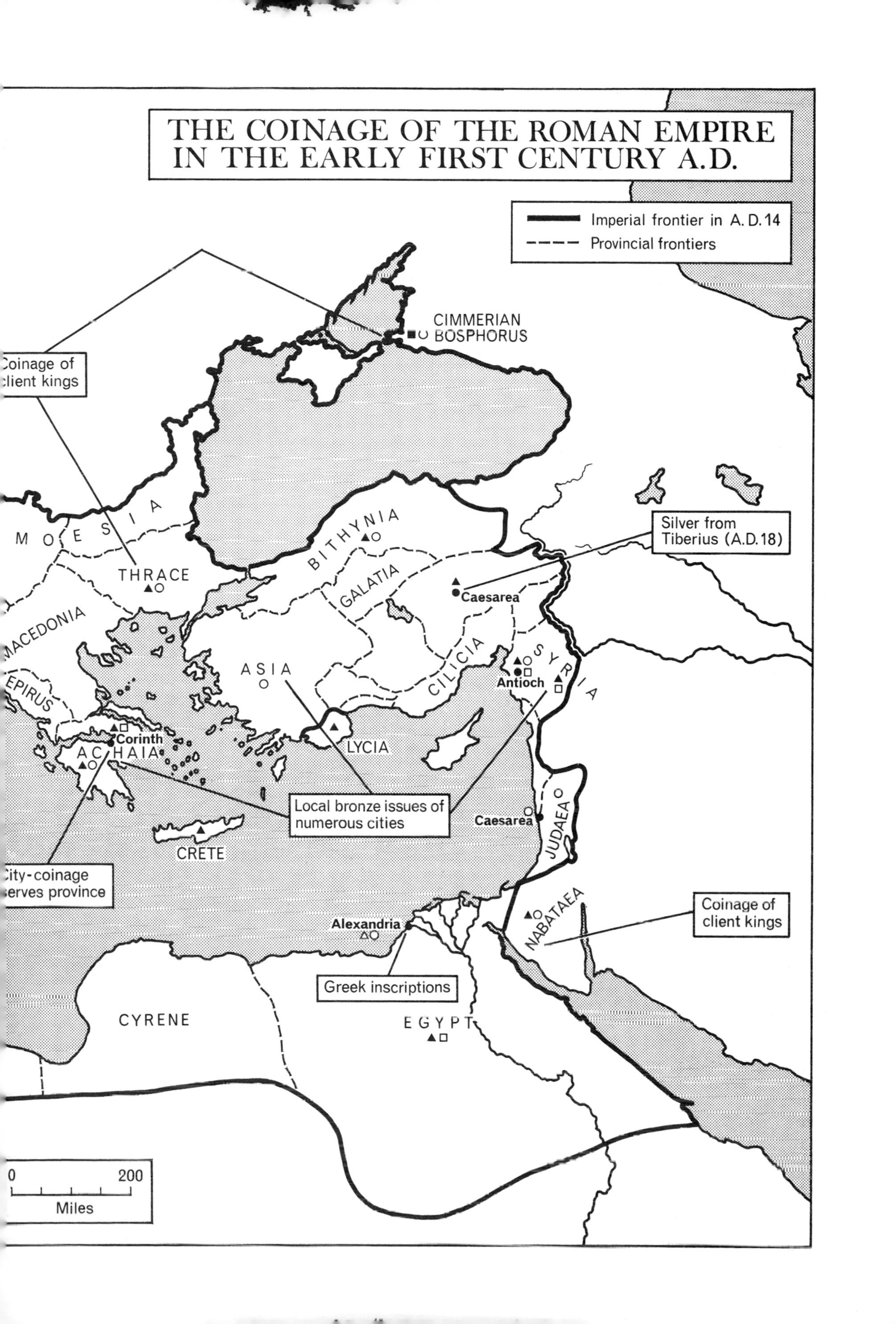

THE COINAGE OF THE ROMAN EMPIRE IN THE EARLY FIRST CENTURY A.D.
Imperial frontier in A. D. 14
Provincial frontiers
CIMMERIAN BOSPHORUS
Coinage of client kings
MOESIA
THRACE
BITHYNIA
GALATIA
Silver from Tiberius (A.D. 18)
Caesarea
MACEDONIA
EPIRUS
ASIA
CILICIA
SYRIA
Antioch
Corinth
ACHAIA
LYCIA
Local bronze issues of numerous cities
Caesarea
JUDAEA
CRETE
City-coinage serves province
NABATAEA
Coinage of client kings
Alexandria
Greek inscriptions
CYRENE
EGYPT
0
200
Miles

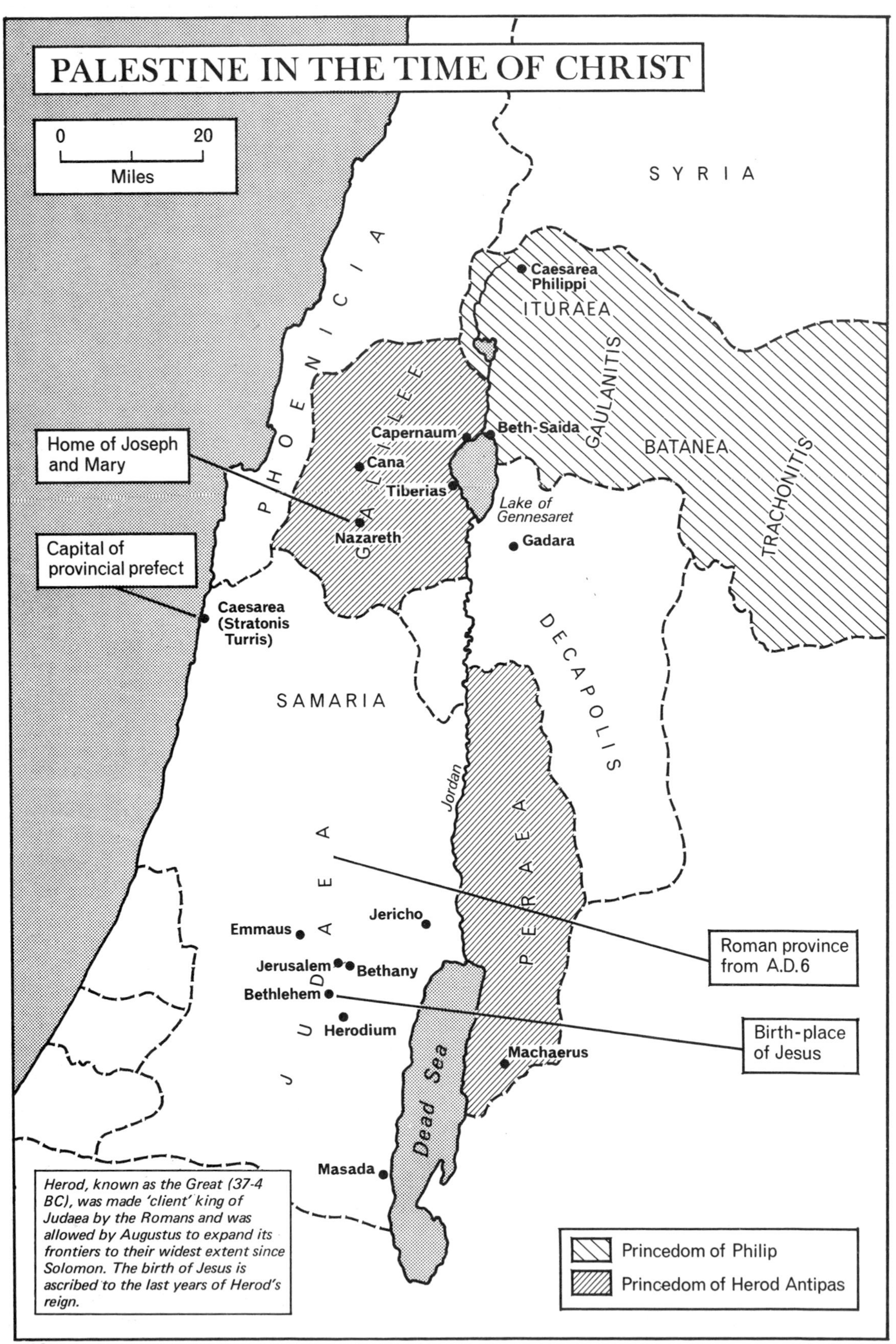

PALESTINE IN THE TIME OF CHRIST
0
20
Miles
SYRIA
PHOENICIA
Caesarea Philippi
ITURAEA
GAULANITIS
BATANEA
TRACHONITIS
GALILEE
Capernaum
Beth-Saida
Cana
Tiberias
Lake of Gennesaret
Nazareth
Gadara
Home of Joseph and Mary
Capital of provincial prefect
Caesarea (Stratonis Turris)
DECAPOLIS
SAMARIA
Jordan
PERAEA
JUDAEA
Jericho
Emmaus
Jerusalem
Bethany
Bethlehem
Herodium
Machaerus
Dead Sea
Masada
Roman province from A.D.6
Birth-place of Jesus
Herod, known as the Great (37-4 BC), was made 'client' king of Judaea by the Romans and was allowed by Augustus to expand its frontiers to their widest extent since Solomon. The birth of Jesus is ascribed to the last years of Herod's reign.
Princedom of Philip
Princedom of Herod Antipas

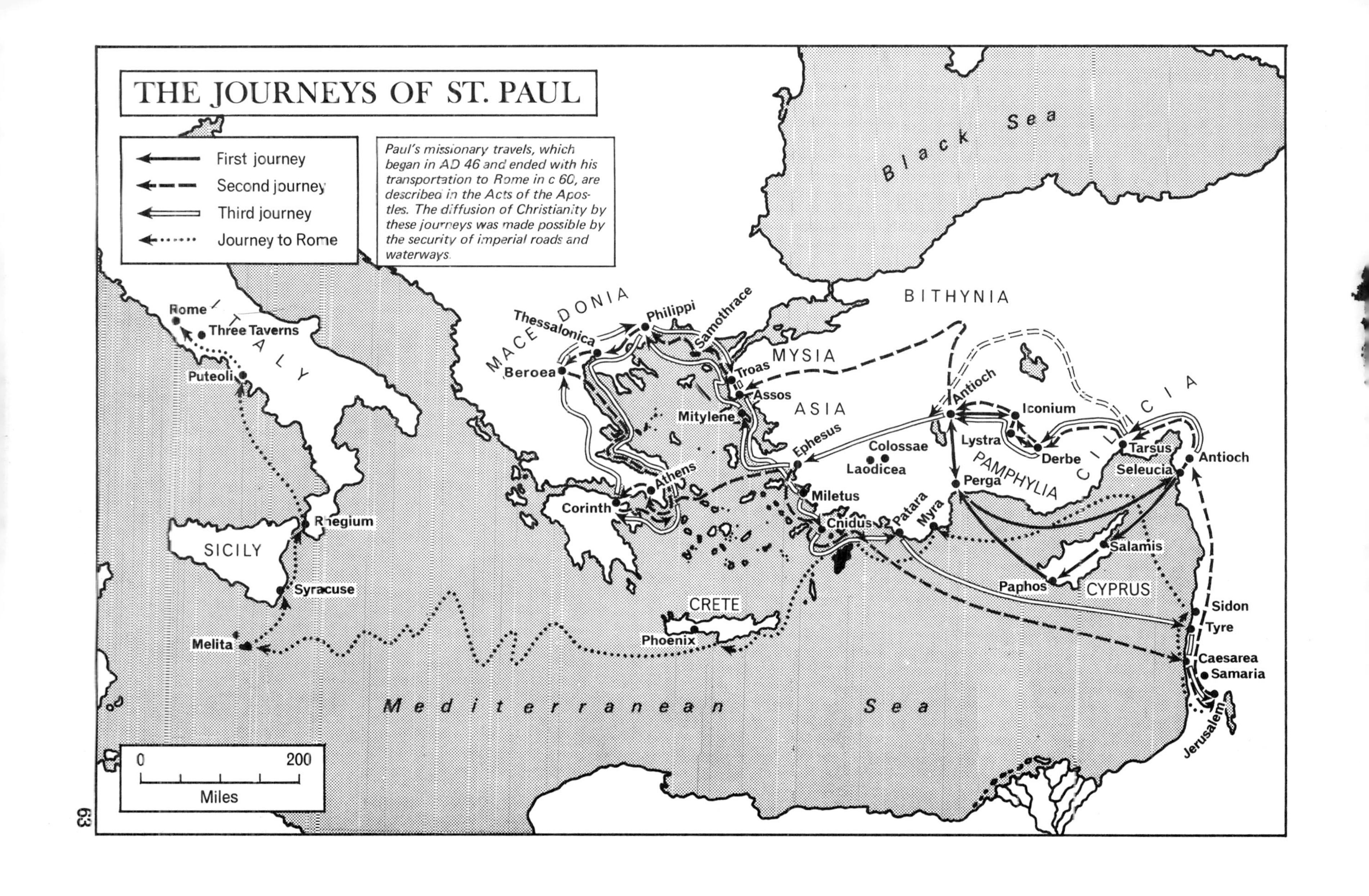
THE JOURNEYS OF ST. PAUL
First journey
Second journey
Third journey
Journey to Rome
Paul's missionary travels, which began in AD 46 and ended with his transportation to Rome in c 60, are described in the Acts of the Apostles. The diffusion of Christianity by these journeys was made possible by the security of imperial roads and waterways.
Black Sea
BITHYNIA
MACEDONIA
MYSIA
ASIA
PAMPHYLIA
CILICIA
CYPRUS
CRETE
SICILY
ITALY
Mediterranean Sea
Rome
Three Taverns
Puteoli
Rhegium
Syracuse
Melita
Thessalonica
Beroea
Philippi
Samothrace
Troas
Assos
Mitylene
Athens
Corinth
Ephesus
Miletus
Cnidus
Patara
Myra
Colossae
Laodicea
Antioch
Iconium
Lystra
Derbe
Perga
Tarsus
Antioch
Seleucia
Salamis
Paphos
Phoenix
Sidon
Tyre
Caesarea
Samaria
Jerusalem
0
200
Miles

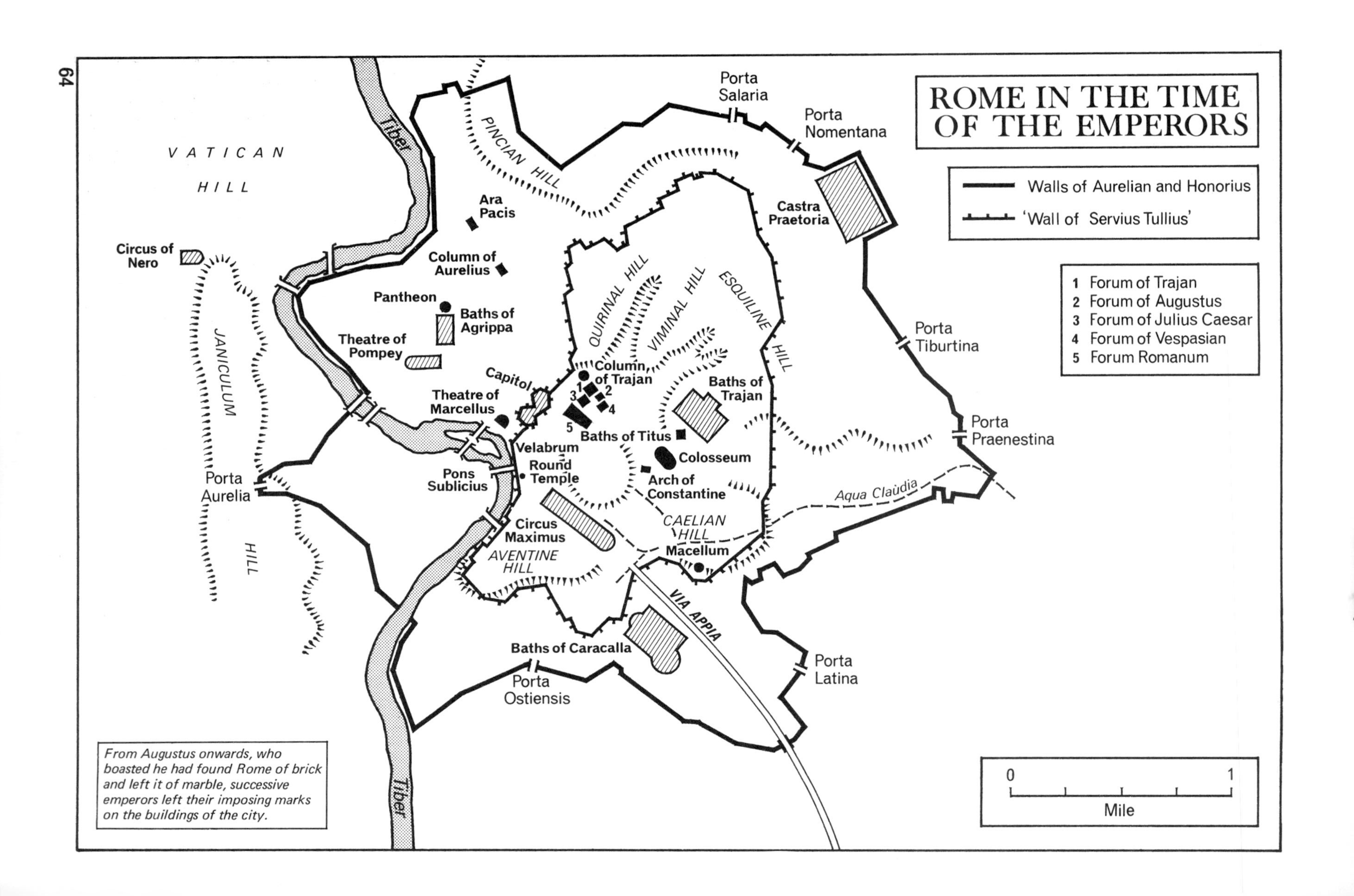

From Augustus onwards, who boasted he had found Rome of brick and left it of marble, successive emperors left their imposing marks on the buildings of the city.

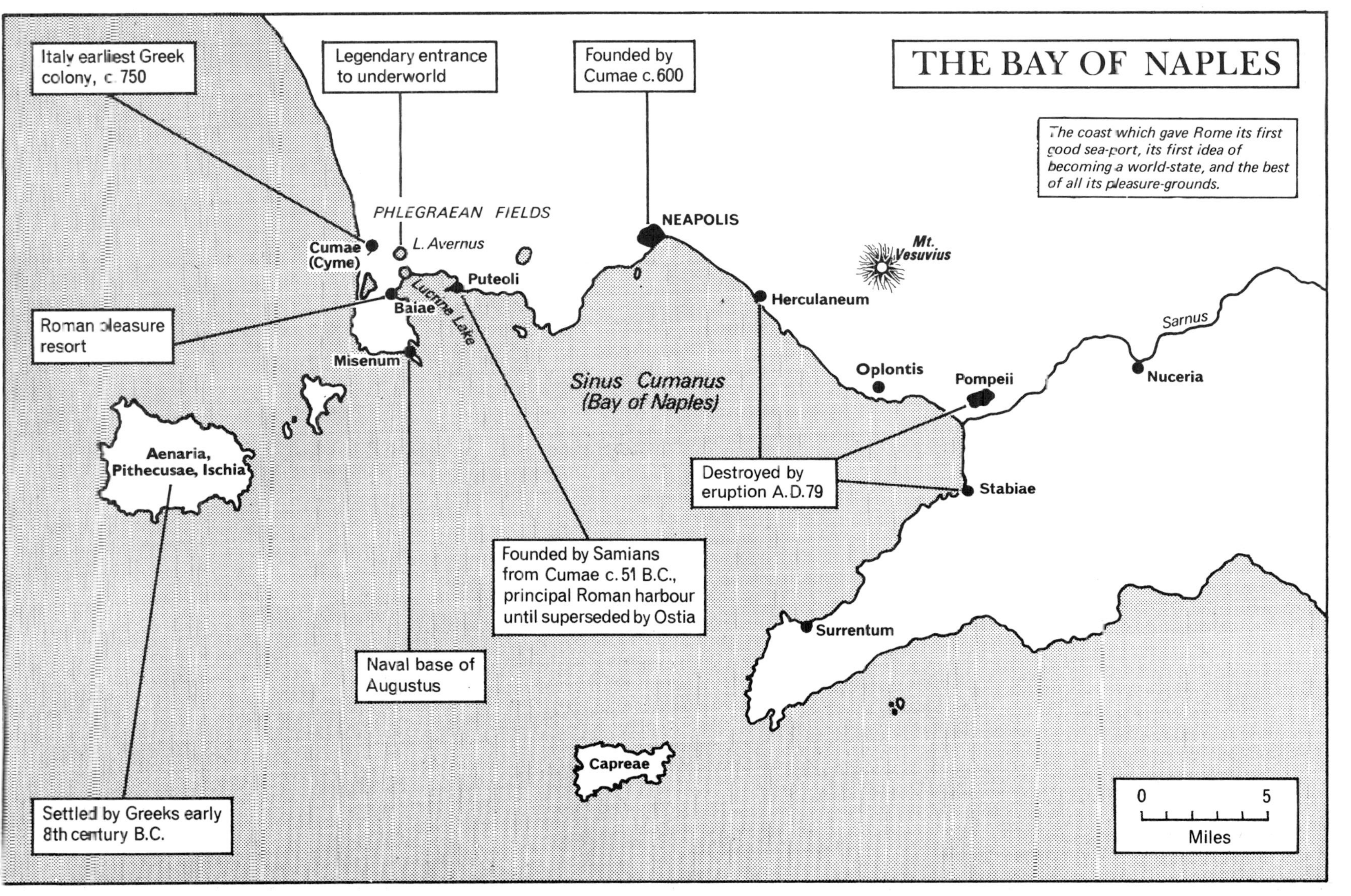

THE BAY OF NAPLES
The coast which gave Rome its first good sea-port, its first idea of becoming a world-state, and the best of all its pleasure-grounds.
Italy earliest Greek colony, c. 750
Legendary entrance to underworld
Founded by Cumae c.600
PHLEGRAEAN FIELDS
NEAPOLIS
Mt. Vesuvius
Cumae
(Cyme)
L. Avernus
Puteoli
Lucrine Lake
Baiae
Misenum
Herculaneum
Roman pleasure resort
Sarnus
Nuceria
Oplontis
Pompeii
Sinus Cumanus
(Bay of Naples)
Aenaria, Pithecusae, Ischia
Destroyed by eruption A.D.79
Stabiae
Founded by Samians from Cumae c.51 B.C., principal Roman harbour until superseded by Ostia
Surrentum
Naval base of Augustus
Capreae
Settled by Greeks early 8th century B.C.
0
5
Miles

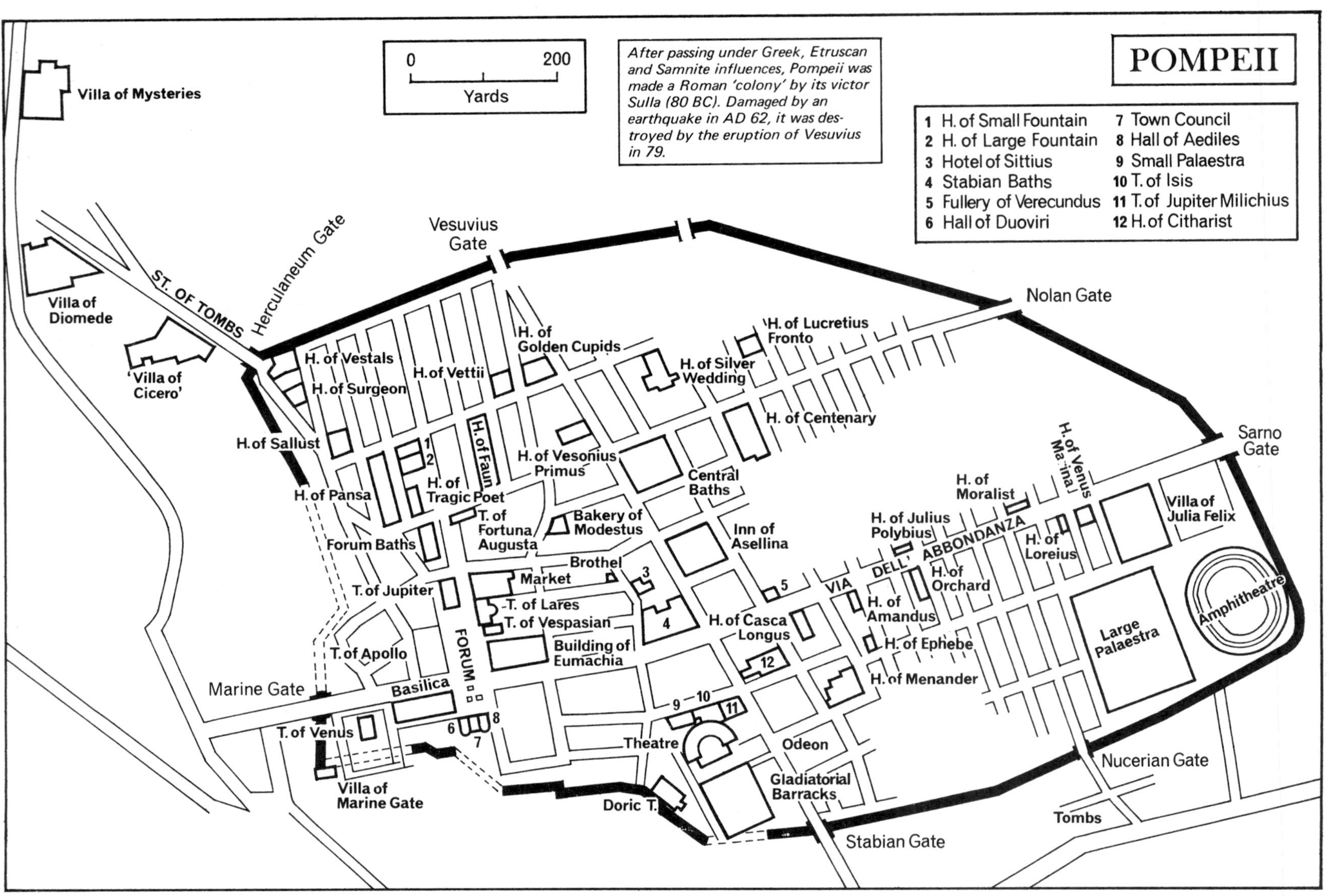
POMPEII
0
200
Yards
After passing under Greek, Etruscan and Samnite influences, Pompeii was made a Roman 'colony' by its victor Sulla (80 BC). Damaged by an earthquake in AD 62, it was destroyed by the eruption of Vesuvius in 79.
1 H. of Small Fountain
2 H. of Large Fountain
3 Hotel of Sittius
4 Stabian Baths
5 Fullery of Verecundus
6 Hall of Duoviri
7 Town Council
8 Hall of Aediles
9 Small Palaestra
10 T. of Isis
11 T. of Jupiter Milichius
12 H. of Citharist
Villa of Mysteries
Villa of Diomede
'Villa of Cicero'
ST. OF TOMBS
Herculaneum Gate
Vesuvius Gate
Nolan Gate
Sarno Gate
Nucerian Gate
Stabian Gate
Marine Gate
Tombs
H. of Vestals
H. of Surgeon
H. of Sallust
H. of Pansa
H. of Vettii
H. of Golden Cupids
H. of Faun
H. of Tragic Poet
H. of Vesonius Primus
H. of Silver Wedding
H. of Lucretius Fronto
H. of Centenary
Central Baths
Inn of Asellina
Bakery of Modestus
T. of Fortuna Augusta
Forum Baths
Brothel
Market
T. of Lares
T. of Vespasian
T. of Jupiter
Building of Eumachia
FORUM
T. of Apollo
Basilica
T. of Venus
Villa of Marine Gate
Doric T.
Theatre
Odeon
Gladiatorial Barracks
H. of Casca Longus
H. of Menander
H. of Ephebe
H. of Amandus
H. of Orchard
VIA DELL' ABBONDANZA
H. of Julius Polybius
H. of Moralist
H. of Loreius
H. of Venus Marina
Villa of Julia Felix
Large Palaestra
Amphitheatre

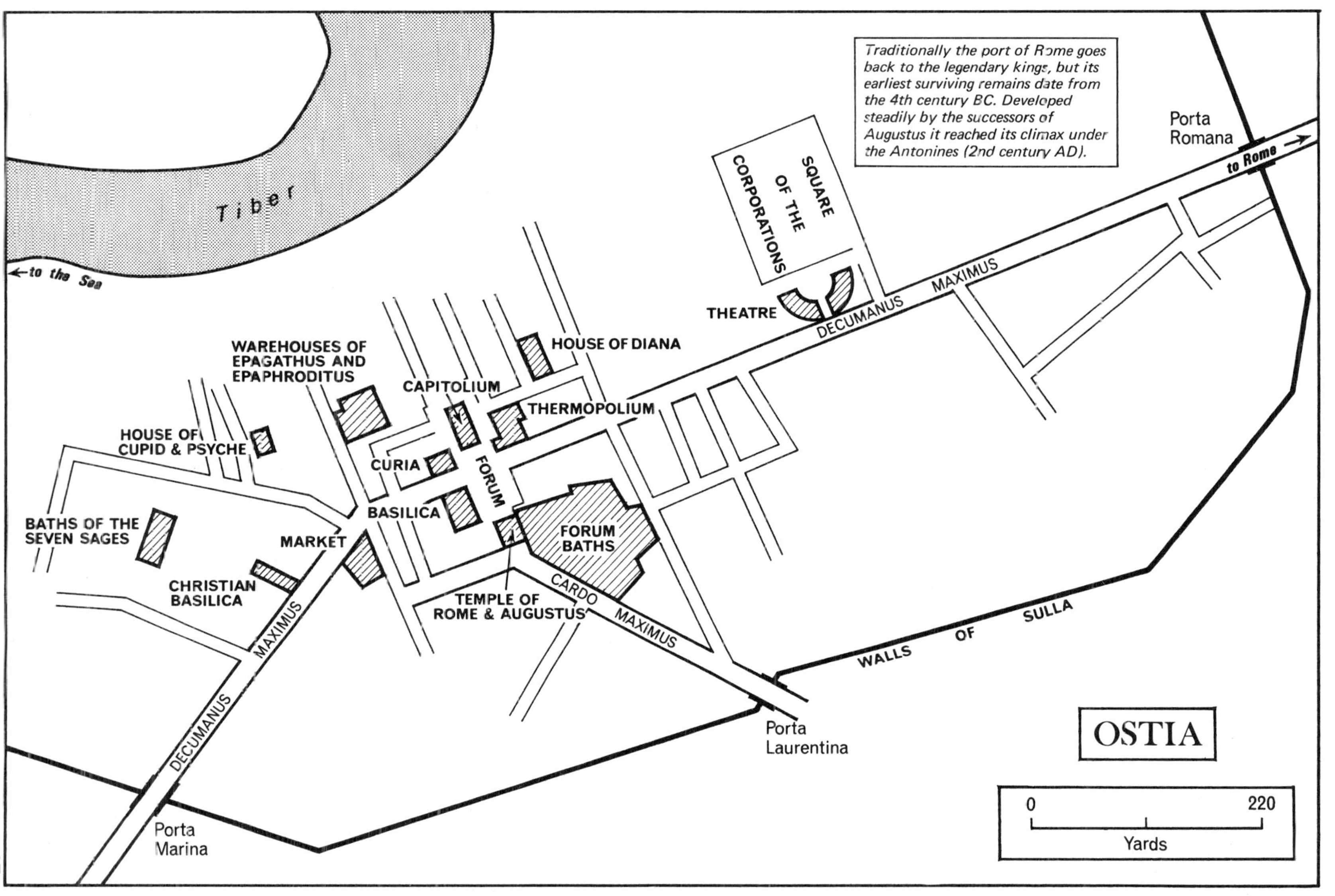
OSTIA
0
220
Yards
Traditionally the port of Rome goes back to the legendary kings, but its earliest surviving remains date from the 4th century BC. Developed steadily by the successors of Augustus it reached its climax under the Antonines (2nd century AD).
Porta Romana
to Rome
Tiber
to the Sea
SQUARE OF THE CORPORATIONS
THEATRE
DECUMANUS MAXIMUS
HOUSE OF DIANA
WAREHOUSES OF EPAGATHUS AND EPAPHRODITUS
CAPITOLIUM
THERMOPOLIUM
HOUSE OF CUPID & PSYCHE
CURIA
FORUM
BASILICA
FORUM BATHS
BATHS OF THE SEVEN SAGES
MARKET
CHRISTIAN BASILICA
TEMPLE OF ROME & AUGUSTUS
CARDO MAXIMUS
WALLS OF SULLA
Porta Laurentina
Porta Marina

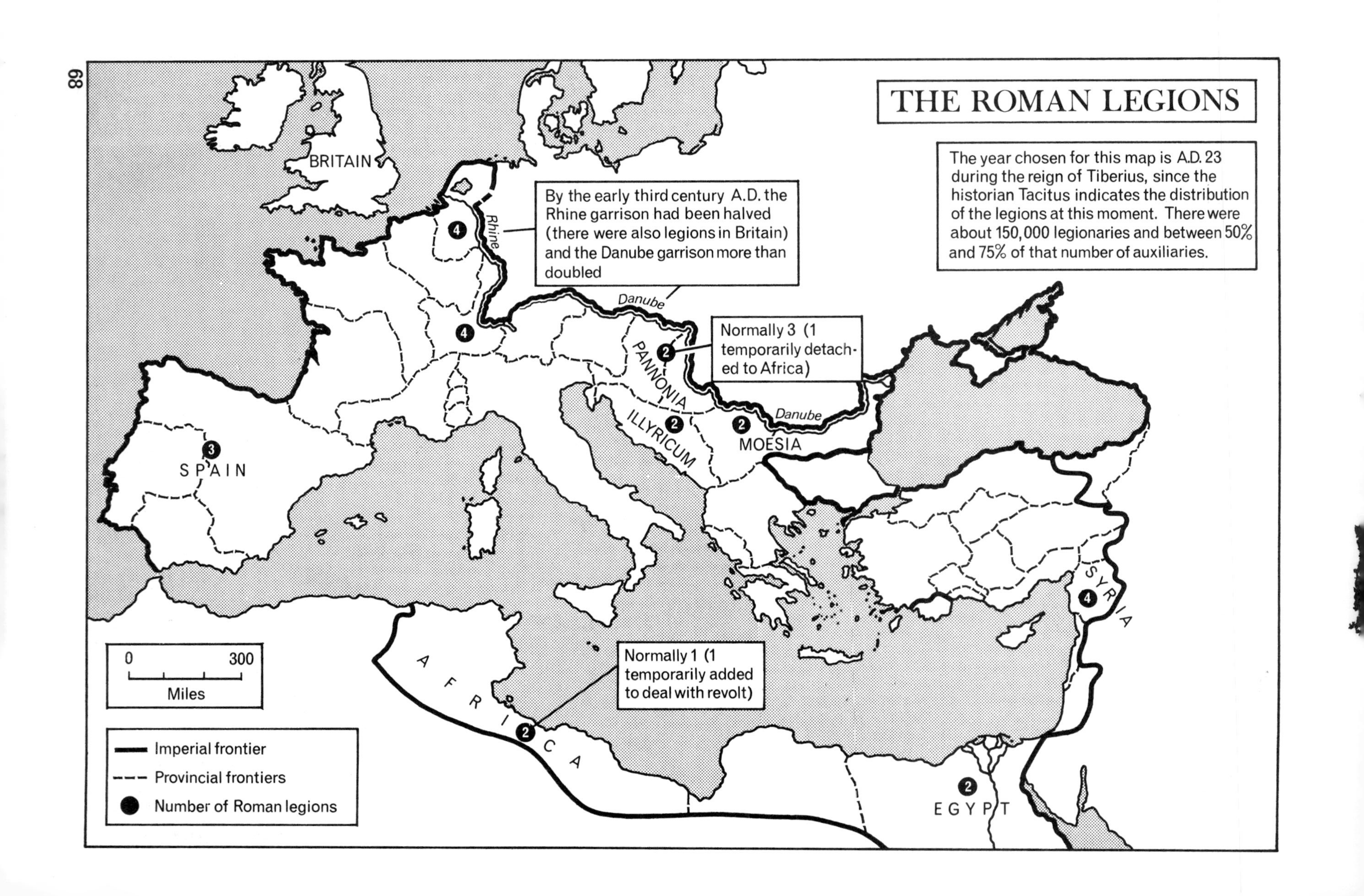
THE ROMAN LEGIONS
The year chosen for this map is A.D. 23 during the reign of Tiberius, since the historian Tacitus indicates the distribution of the legions at this moment. There were about 150,000 legionaries and between 50% and 75% of that number of auxiliaries.
By the early third century A.D. the Rhine garrison had been halved (there were also legions in Britain) and the Danube garrison more than doubled
Normally 3 (1 temporarily detach-ed to Africa)
Normally 1 (1 temporarily added to deal with revolt)
BRITAIN
Rhine
Danube
Danube
4
4
PANNONIA
2
ILLYRICUM
2
MOESIA
2
SPAIN
3
SYRIA
4
AFRICA
2
EGYPT
2
0
300
Miles
Imperial frontier
Provincial frontiers
Number of Roman legions

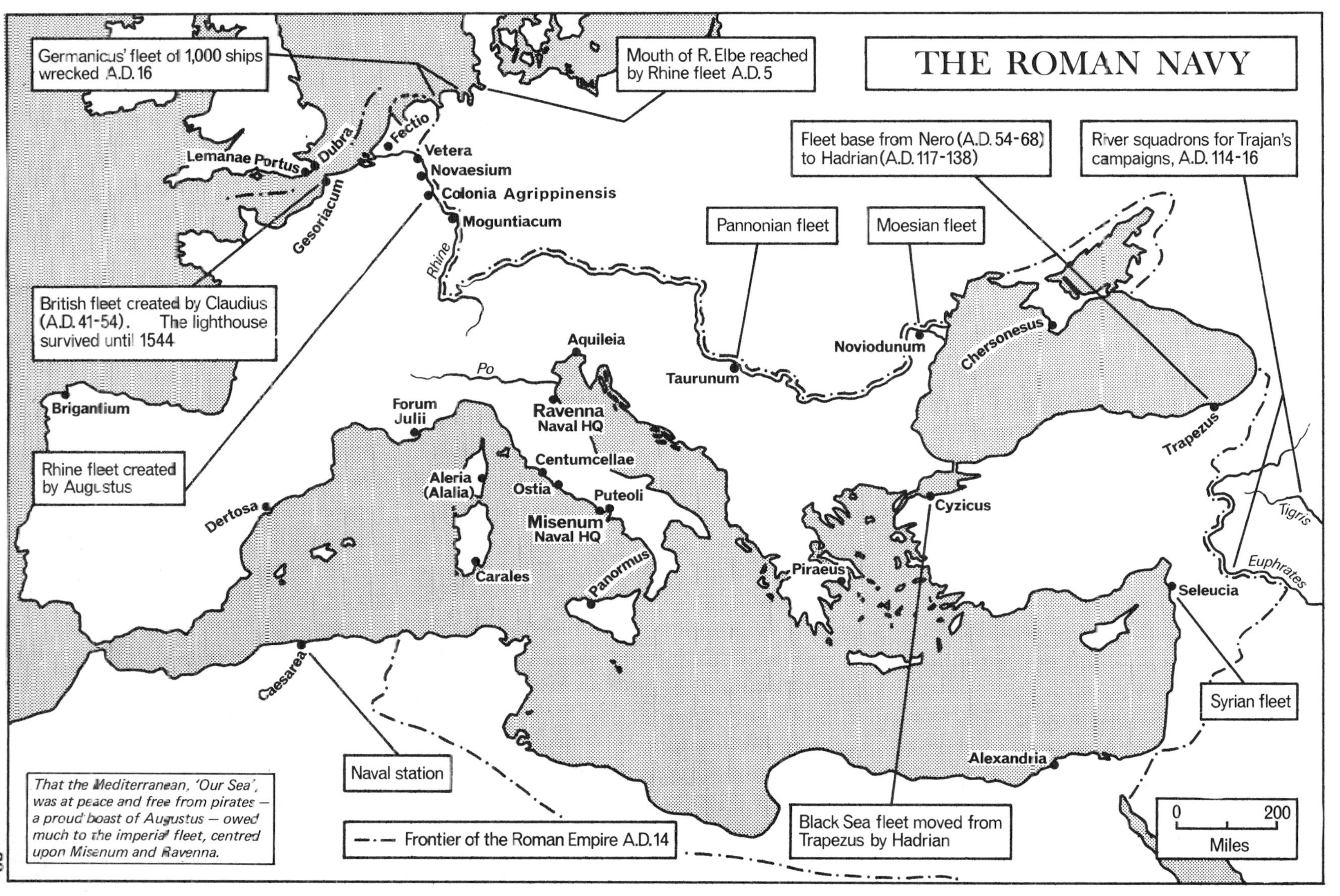
THE ROMAN NAVY
Germanicus' fleet of 1,000 ships wrecked A.D.16
Mouth of R. Elbe reached by Rhine fleet A.D.5
Fleet base from Nero (A.D.54-68) to Hadrian (A.D.117-138)
River squadrons for Trajan's campaigns, A.D.114-16
Pannonian fleet
Moesian fleet
British fleet created by Claudius (A.D.41-54). The lighthouse survived until 1544
Rhine fleet created by Augustus
Syrian fleet
Naval station
Black Sea fleet moved from Trapezus by Hadrian
Frontier of the Roman Empire A.D.14
0
200
Miles
That the Mediterranean, 'Our Sea', was at peace and free from pirates – a proud boast of Augustus – owed much to the imperial fleet, centred upon Misenum and Ravenna.
Lemanae Portus
Dubra
Fectio
Gesoriacum
Vetera
Novaesium
Colonia Agrippinensis
Moguntiacum
Rhine
Brigantium
Po
Aquileia
Ravenna
Naval HQ
Forum Julii
Aleria (Alalia)
Centumcellae
Ostia
Puteoli
Misenum
Naval HQ
Carales
Dertosa
Panormus
Caesarea
Taurunum
Noviodunum
Chersonesus
Trapezus
Cyzicus
Piraeus
Tigris
Euphrates
Seleucia
Alexandria

FREE GERMANY
BRITANNIA
(AD 71)
(AD 59)
(AD 43-47)
Londinium
LOWER GERMANY
Colonia Agrippinensis
Moguntiacum
Rhine
AGRI DECUMAT. (83)
LUGDUNENSIS
UPPER GERMANY
RHAETIA
NORICUM
Danube
PANNONIA
UPPER
LOWER
GALLIA
Lugdunum
AQUITANIA
NARBONENSIS
Aquileia
ITALIA
ILLYRICUM
Adriatic Sea
Nemausus
TARRACONENSIS
Tarraco
Rome
HISPANIA
LUSITANIA
SARDINIA
BAETICA
Corduba
Gades
SICILY
Carthage
MAURETANIA (A.D. 42)
AFRICA
Frontier of Roman Empire A.D. 14
Frontier of Roman Empire A.D. 117
Province boundaries

THE ROMAN EMPIRE FROM TIBERIUS (A.D. 14-37) TO TRAJAN (98-117)

Trajan's expansion as far as the Persian Gulf came to nothing, since his successor Hadrian withdrew to the Euphrates again.

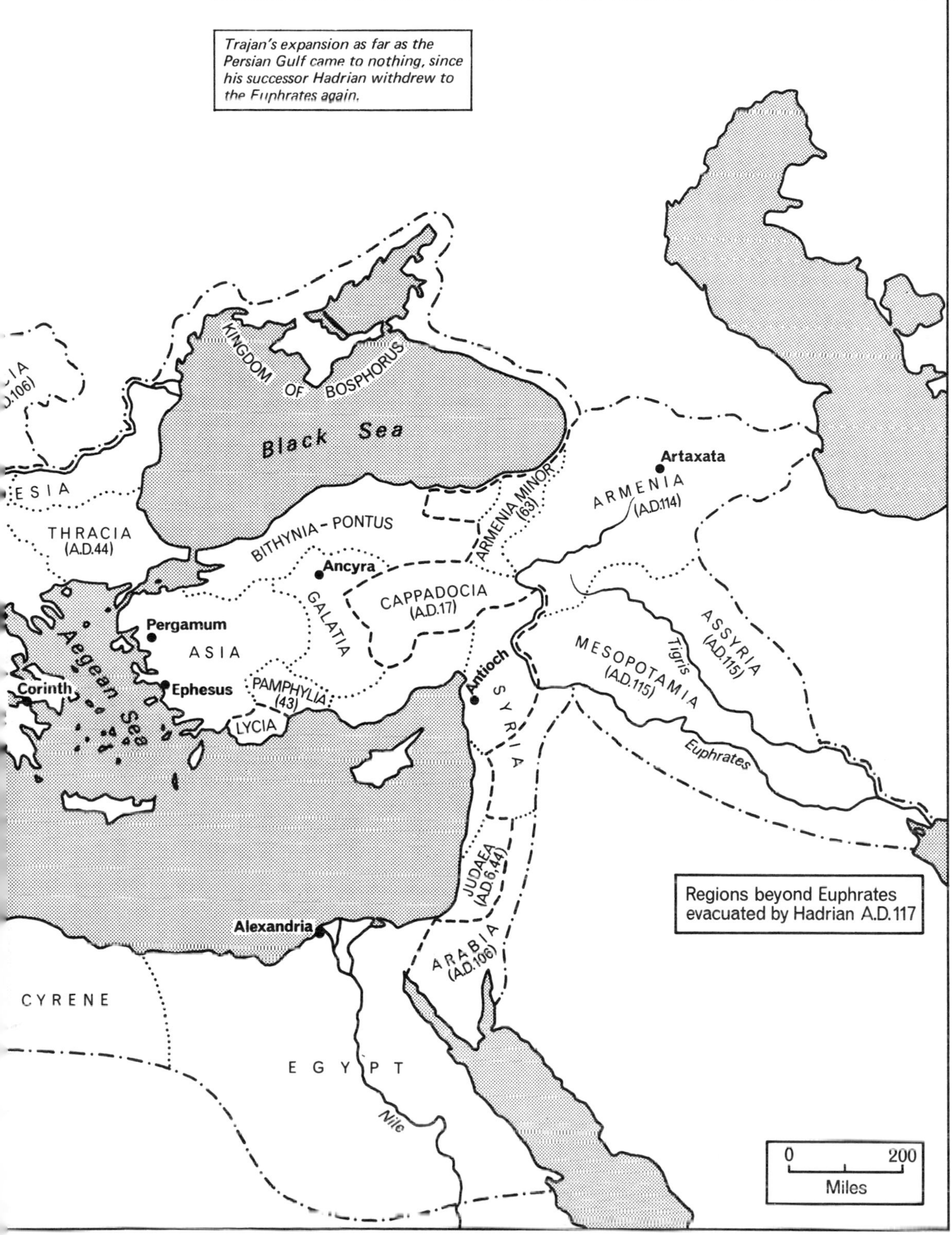

KINGDOM OF BOSPHORUS
Black Sea
Artaxata
ARMENIA (A.D.114)
ARMENIA MINOR (63)
BITHYNIA - PONTUS
THRACIA (A.D.44)
ESIA
IA (A.D.106)
Ancyra
GALATIA
CAPPADOCIA (A.D.17)
Pergamum
ASIA
Aegean Sea
Corinth
Ephesus
PAMPHYLIA (43)
LYCIA
Antioch
SYRIA
MESOPOTAMIA (A.D.115)
Tigris
ASSYRIA (A.D.115)
Euphrates
JUDAEA (A.D.6,44)
ARABIA (A.D.106)
Alexandria
CYRENE
EGYPT
Nile

Regions beyond Euphrates evacuated by Hadrian A.D. 117

0 200
Miles

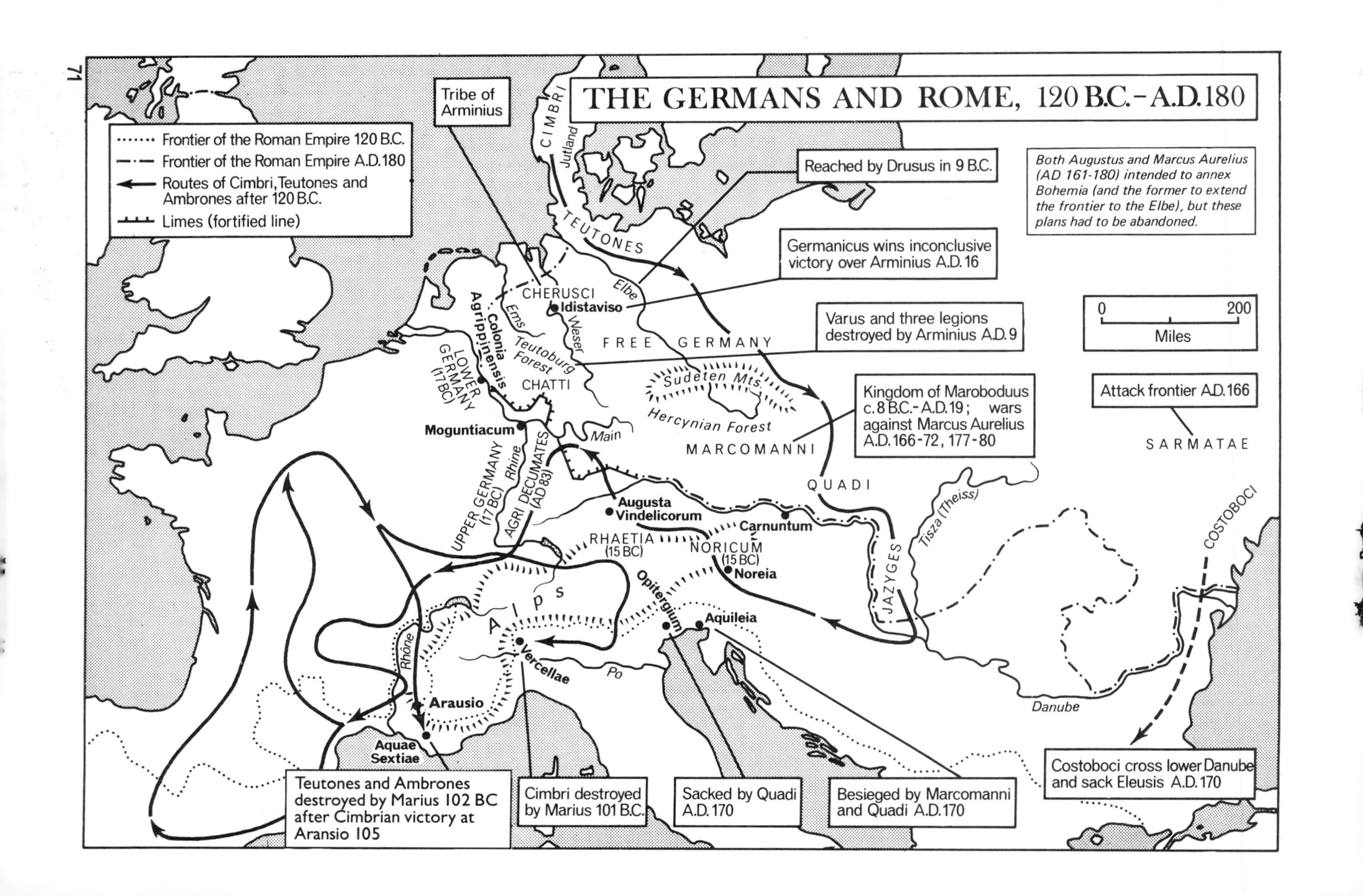

THE GERMANS AND ROME, 120 B.C.-A.D.180
Frontier of the Roman Empire 120 B.C.
Frontier of the Roman Empire A.D.180
Routes of Cimbri, Teutones and Ambrones after 120 B.C.
Limes (fortified line)
Tribe of Arminius
Reached by Drusus in 9 B.C.
Both Augustus and Marcus Aurelius (AD 161-180) intended to annex Bohemia (and the former to extend the frontier to the Elbe), but these plans had to be abandoned.
Germanicus wins inconclusive victory over Arminius A.D.16
Varus and three legions destroyed by Arminius A.D.9
Kingdom of Maroboduus c.8 B.C.-A.D.19; wars against Marcus Aurelius A.D.166-72, 177-80
0
200
Miles
Attack frontier A.D.166
SARMATAE
CIMBRI
Jutland
TEUTONES
Elbe
CHERUSCI
Idistaviso
Ems
Weser
Teutoburg Forest
FREE GERMANY
Sudeten Mts.
Hercynian Forest
MARCOMANNI
QUADI
CHATTI
Colonia Agrippinensis
LOWER GERMANY (17 BC)
Moguntiacum
Main
UPPER GERMANY (17 BC)
Rhine
AGRI DECUMATES (AD 83)
Augusta Vindelicorum
Carnuntum
RHAETIA (15 BC)
NORICUM (15 BC)
Noreia
JAZYGES
Tisza (Theiss)
COSTOBOCI
Opitergium
Aquileia
Alps
Rhône
Vercellae
Po
Danube
Arausio
Aquae Sextiae
Teutones and Ambrones destroyed by Marius 102 BC after Cimbrian victory at Aransio 105
Cimbri destroyed by Marius 101 B.C.
Sacked by Quadi A.D.170
Besieged by Marcomanni and Quadi A.D.170
Costoboci cross lower Danube and sack Eleusis A.D.170

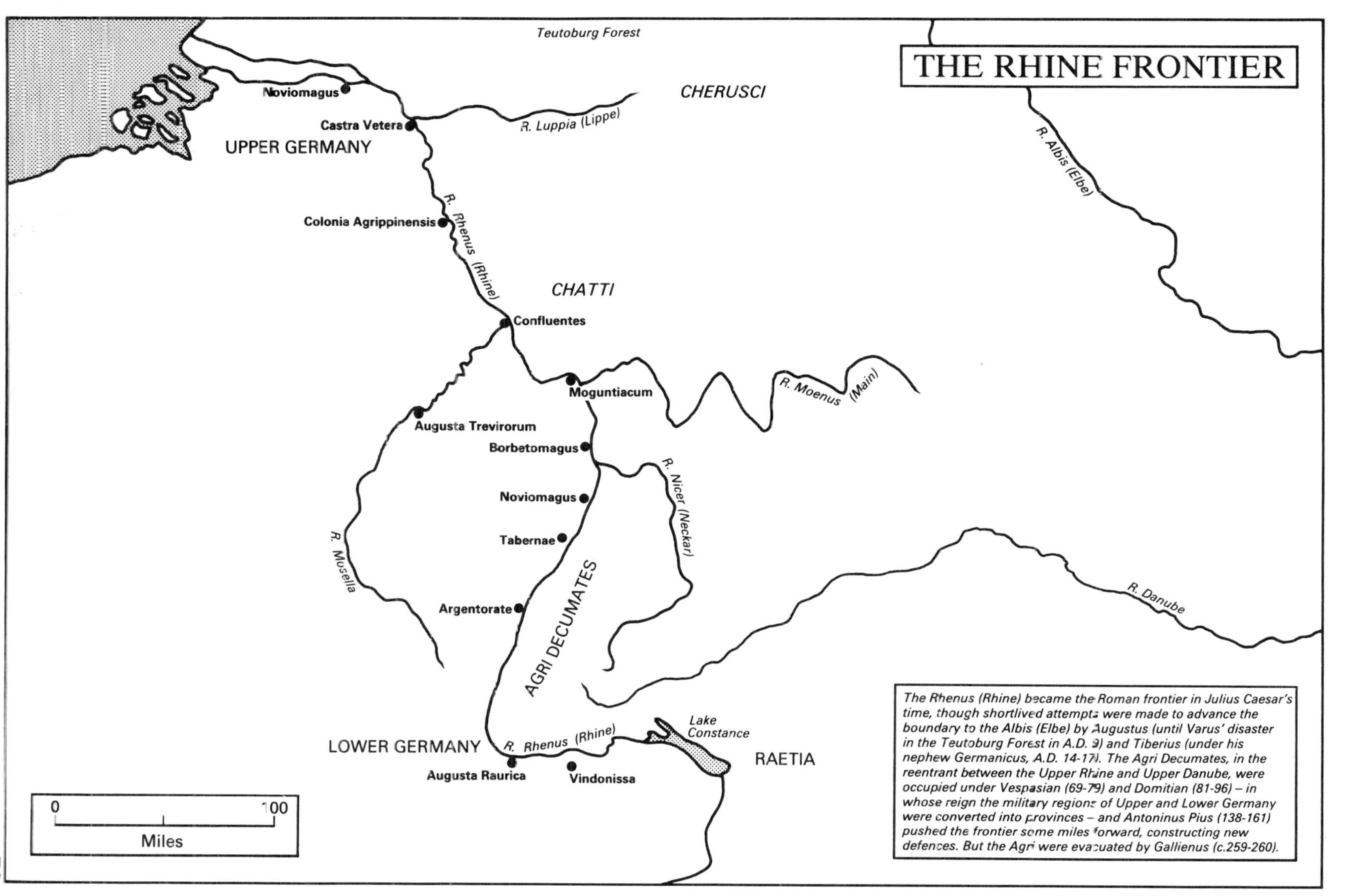

The Rhenus (Rhine) became the Roman frontier in Julius Caesar's time, though shortlived attempts were made to advance the boundary to the Albis (Elbe) by Augustus (until Varus' disaster in the Teutoburg Forest in A.D. 9) and Tiberius (under his nephew Germanicus, A.D. 14-17). The Agri Decumates, in the reentrant between the Upper Rhine and Upper Danube, were occupied under Vespasian (69-79) and Domitian (81-96) – in whose reign the military regions of Upper and Lower Germany were converted into provinces – and Antoninus Pius (138-161) pushed the frontier some miles forward, constructing new defences. But the Agri were evacuated by Gallienus (c.259-260).

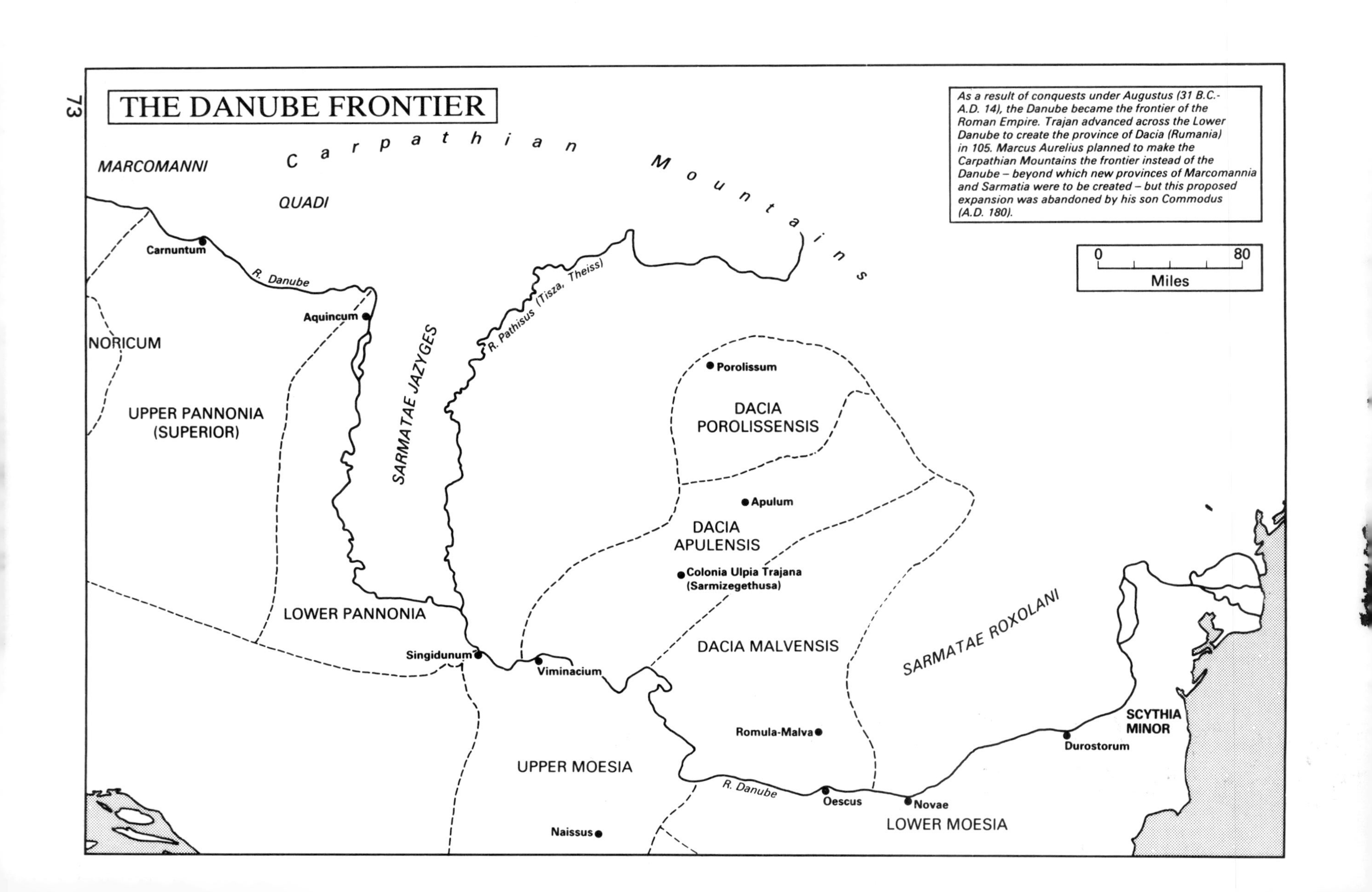
THE DANUBE FRONTIER
As a result of conquests under Augustus (31 B.C.-A.D. 14), the Danube became the frontier of the Roman Empire. Trajan advanced across the Lower Danube to create the province of Dacia (Rumania) in 105. Marcus Aurelius planned to make the Carpathian Mountains the frontier instead of the Danube – beyond which new provinces of Marcomannia and Sarmatia were to be created – but this proposed expansion was abandoned by his son Commodus (A.D. 180).
0
80
Miles
MARCOMANNI
Carpathian Mountains
QUADI
Carnuntum
R. Danube
Aquincum
NORICUM
UPPER PANNONIA (SUPERIOR)
SARMATAE JAZYGES
R. Pathisus (Tisza, Theiss)
Porolissum
DACIA POROLISSENSIS
Apulum
DACIA APULENSIS
Colonia Ulpia Trajana (Sarmizegethusa)
LOWER PANNONIA
Singidunum
Viminacium
DACIA MALVENSIS
SARMATAE ROXOLANI
SCYTHIA MINOR
Durostorum
Romula-Malva
UPPER MOESIA
R. Danube
Oescus
Novae
LOWER MOESIA
Naissus

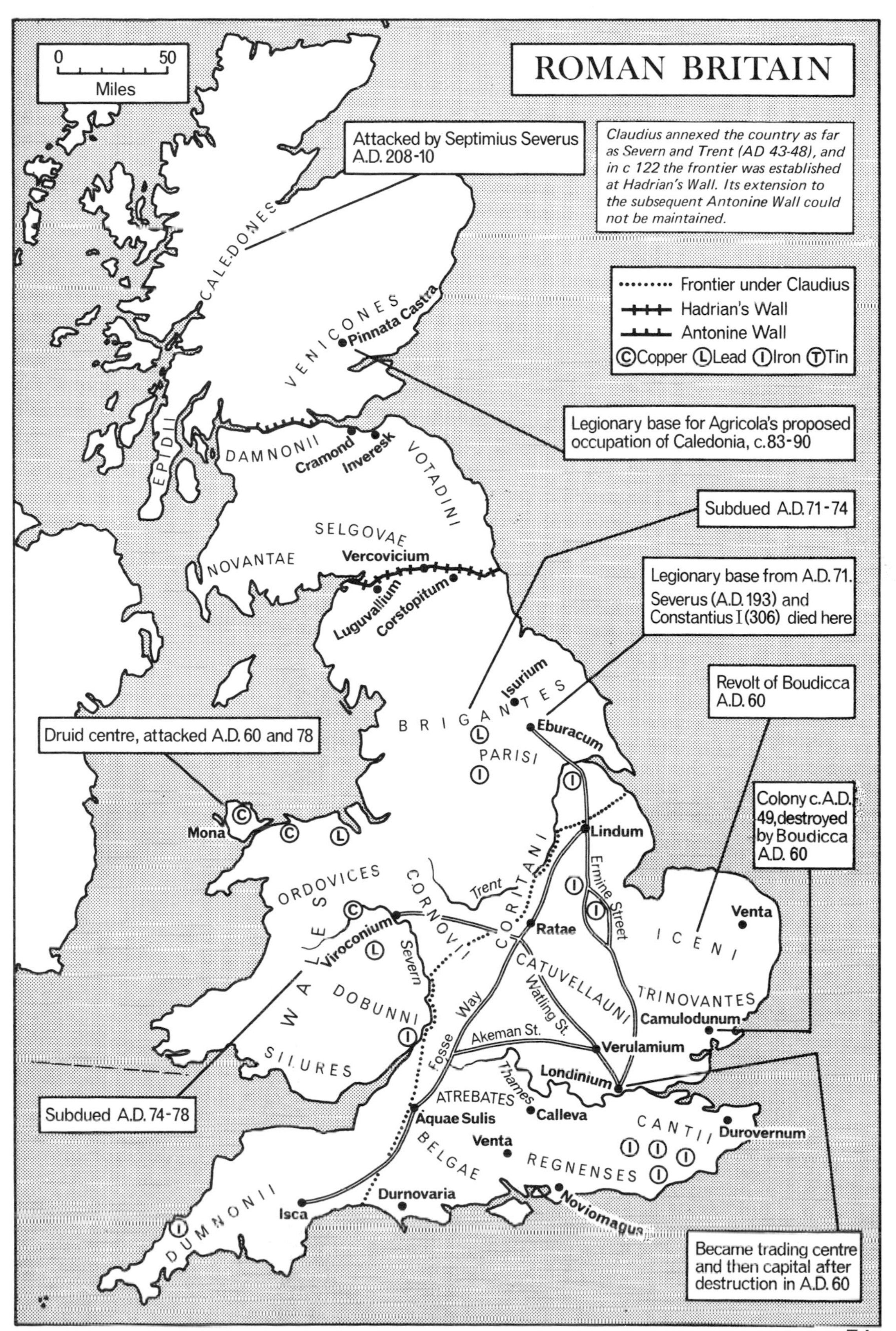
ROMAN BRITAIN
0 50
Miles
Attacked by Septimius Severus A.D. 208-10
Claudius annexed the country as far as Severn and Trent (AD 43-48), and in c 122 the frontier was established at Hadrian's Wall. Its extension to the subsequent Antonine Wall could not be maintained.
Frontier under Claudius
Hadrian's Wall
Antonine Wall
Copper Lead Iron Tin
Legionary base for Agricola's proposed occupation of Caledonia, c.83-90
Subdued A.D. 71-74
Legionary base from A.D. 71. Severus (A.D. 193) and Constantius I (306) died here
Revolt of Boudicca A.D. 60
Druid centre, attacked A.D. 60 and 78
Colony c. A.D. 49, destroyed by Boudicca A.D. 60
Subdued A.D. 74-78
Became trading centre and then capital after destruction in A.D. 60
CALEDONES
VENICONES
Pinnata Castra
EPIDII
DAMNONII
Cramond
Inveresk
VOTADINI
SELGOVAE
NOVANTAE
Vercovicium
Luguvallium
Corstopitum
Isurium
BRIGANTES
Eburacum
PARISI
Mona
Lindum
Trent
ORDOVICES
CORNOVII
CORITANI
Ermine Street
Venta
ICENI
Ratae
Viroconium
WALES
Severn
CATUVELLAUNI
Watling St.
DOBUNNI
TRINOVANTES
Camulodunum
Fosse Way
Akeman St.
Verulamium
SILURES
Thames
Londinium
ATREBATES
Aquae Sulis
Calleva
CANTII
Durovernum
BELGAE
Venta
REGNENSES
Durnovaria
Noviomagus
DUMNONII
Isca

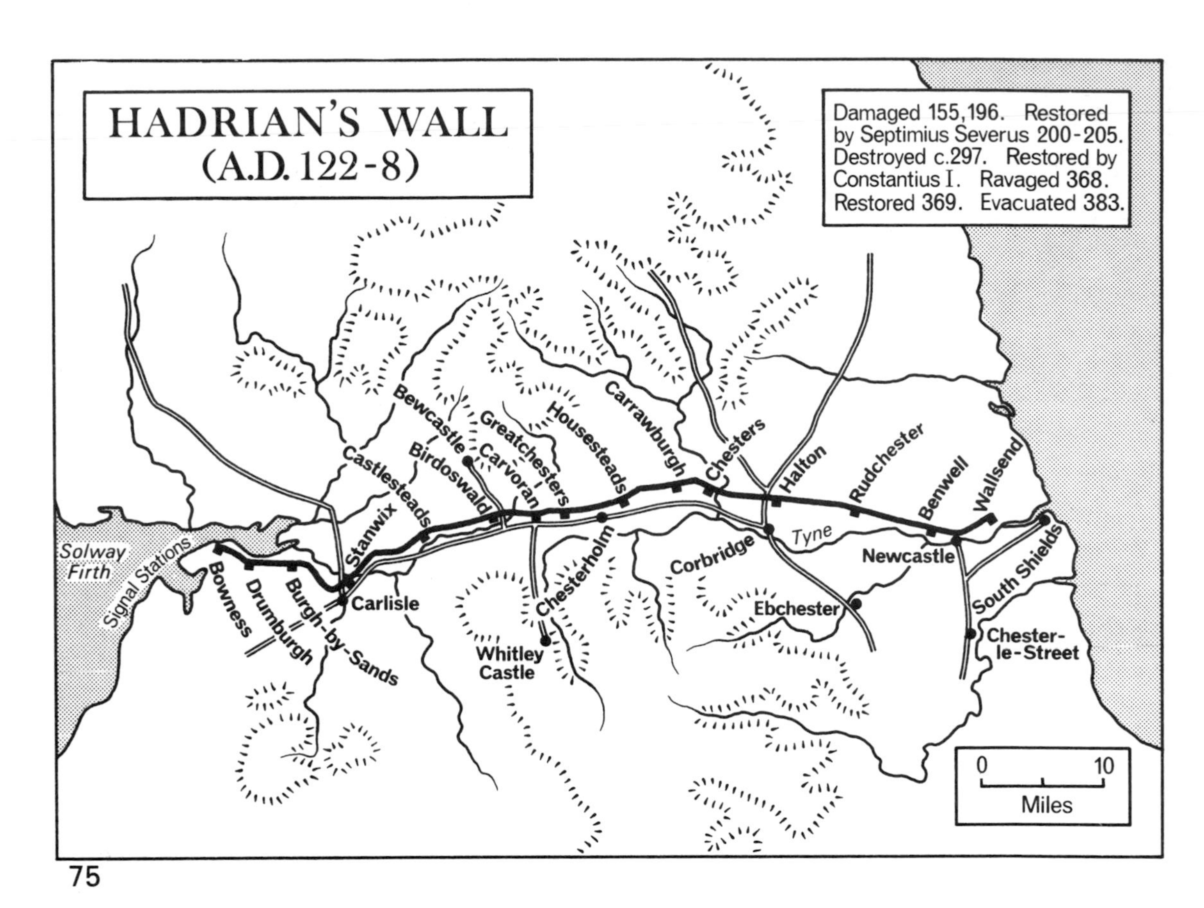
HADRIAN'S WALL
(A.D. 122-8)
Damaged 155,196. Restored by Septimius Severus 200-205. Destroyed c.297. Restored by Constantius I. Ravaged 368. Restored 369. Evacuated 383.
Solway Firth
Signal Stations
Bowness
Drumburgh
Burgh-by-Sands
Carlisle
Stanwix
Castlesteads
Bewcastle
Birdoswald
Greatchesters
Carvoran
Housesteads
Carrawburgh
Chesters
Halton
Rudchester
Benwell
Wallsend
Chesterholm
Whitley Castle
Corbridge
Tyne
Newcastle
Ebchester
South Shields
Chester-le-Street
0
10
Miles

75

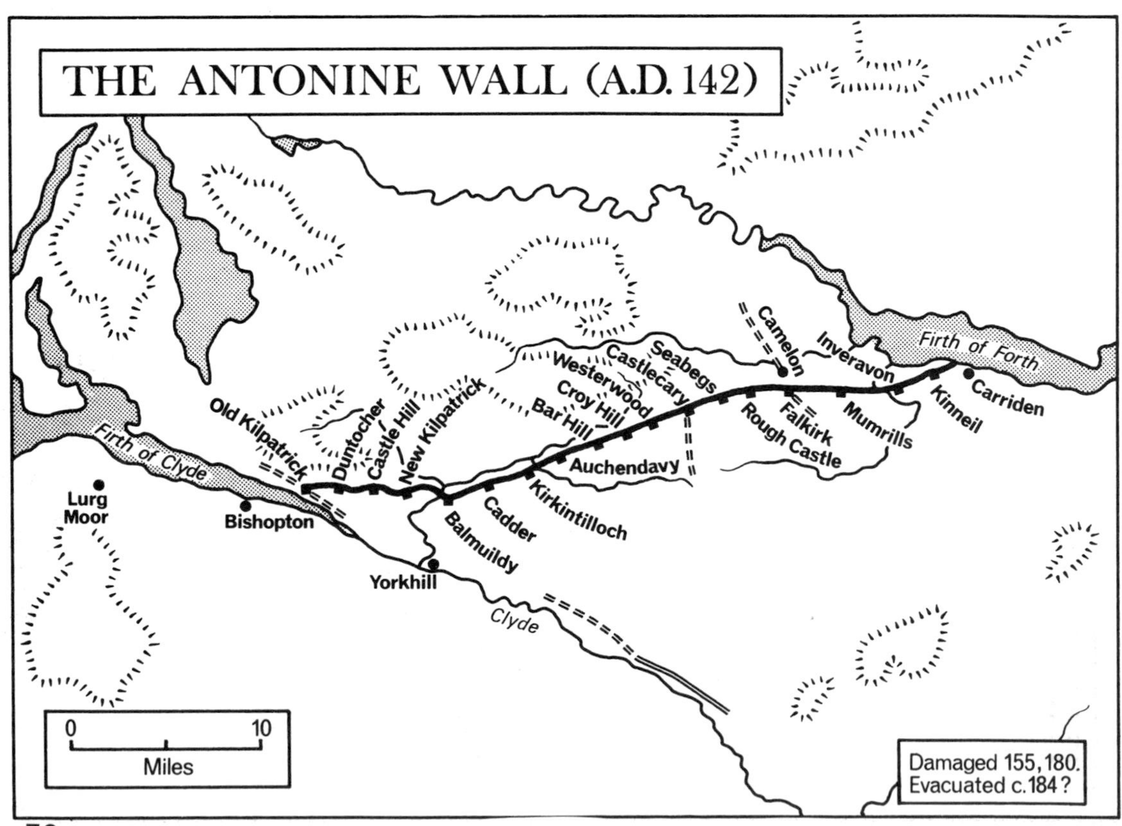
THE ANTONINE WALL (A.D. 142)
Firth of Forth
Camelon
Inveravon
Carriden
Kinneil
Mumrills
Falkirk
Rough Castle
Seabegs
Castlecary
Westerwood
Croy Hill
Bar Hill
Auchendavy
Kirkintilloch
Cadder
Balmuildy
New Kilpatrick
Castle Hill
Duntocher
Old Kilpatrick
Firth of Clyde
Lurg Moor
Bishopton
Yorkhill
Clyde
0
10
Miles
Damaged 155,180. Evacuated c.184?

76

THE WORLD ACCORDING TO PTOLEMY, c. A.D. 150

The Geography of Claudius Ptolemaeus of Alexandria, including an atlas, showed awareness of the existence of China, but not of its shape.

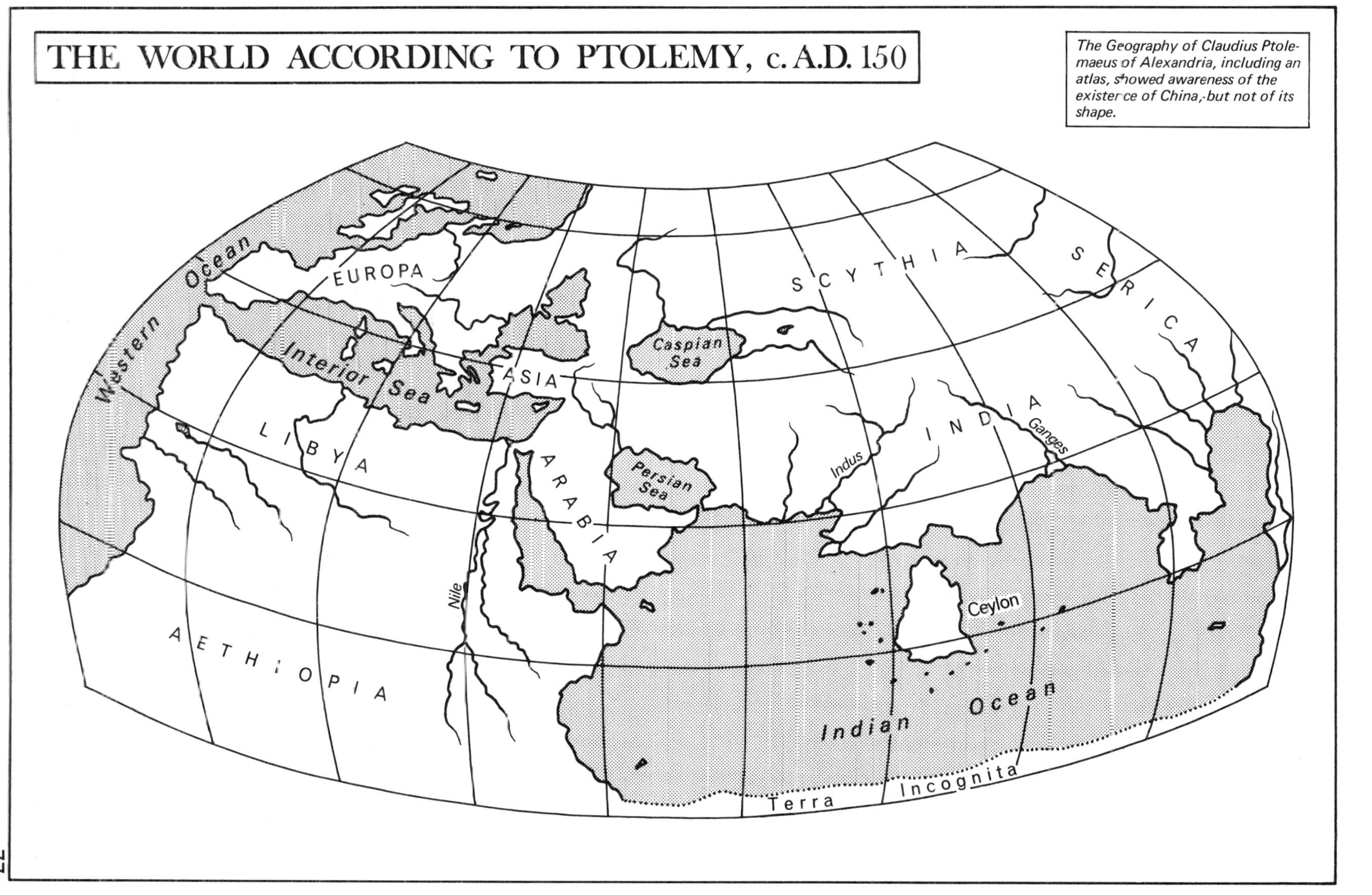

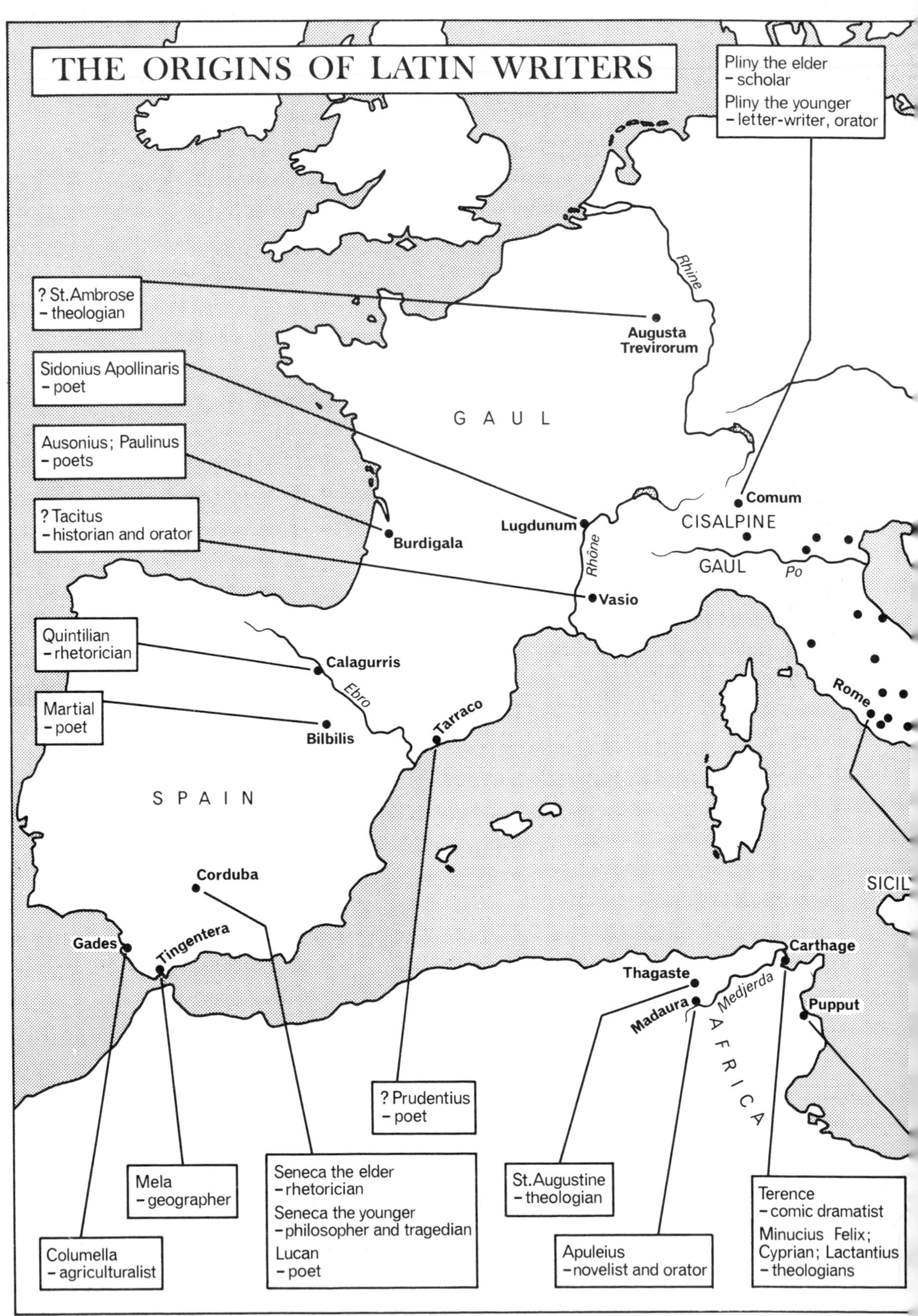
THE ORIGINS OF LATIN WRITERS
Pliny the elder
- scholar
Pliny the younger
- letter-writer, orator
? St.Ambrose
- theologian
Sidonius Apollinaris
- poet
Ausonius; Paulinus
- poets
? Tacitus
- historian and orator
Quintilian
- rhetorician
Martial
- poet
Columella
- agriculturalist
Mela
- geographer
Seneca the elder
- rhetorician
Seneca the younger
- philosopher and tragedian
Lucan
- poet
? Prudentius
- poet
St.Augustine
- theologian
Apuleius
- novelist and orator
Terence
- comic dramatist
Minucius Felix;
Cyprian; Lactantius
- theologians
Rhine
Augusta
Treverorum
G A U L
Comum
CISALPINE
GAUL
Po
Lugdunum
Rhône
Burdigala
Vasio
Calagurris
Ebro
Bilbilis
Tarraco
Rome
S P A I N
Corduba
Gades
Tingentera
SICIL
Carthage
Thagaste
Madaura
Medjerda
Pupput
A F R I C A

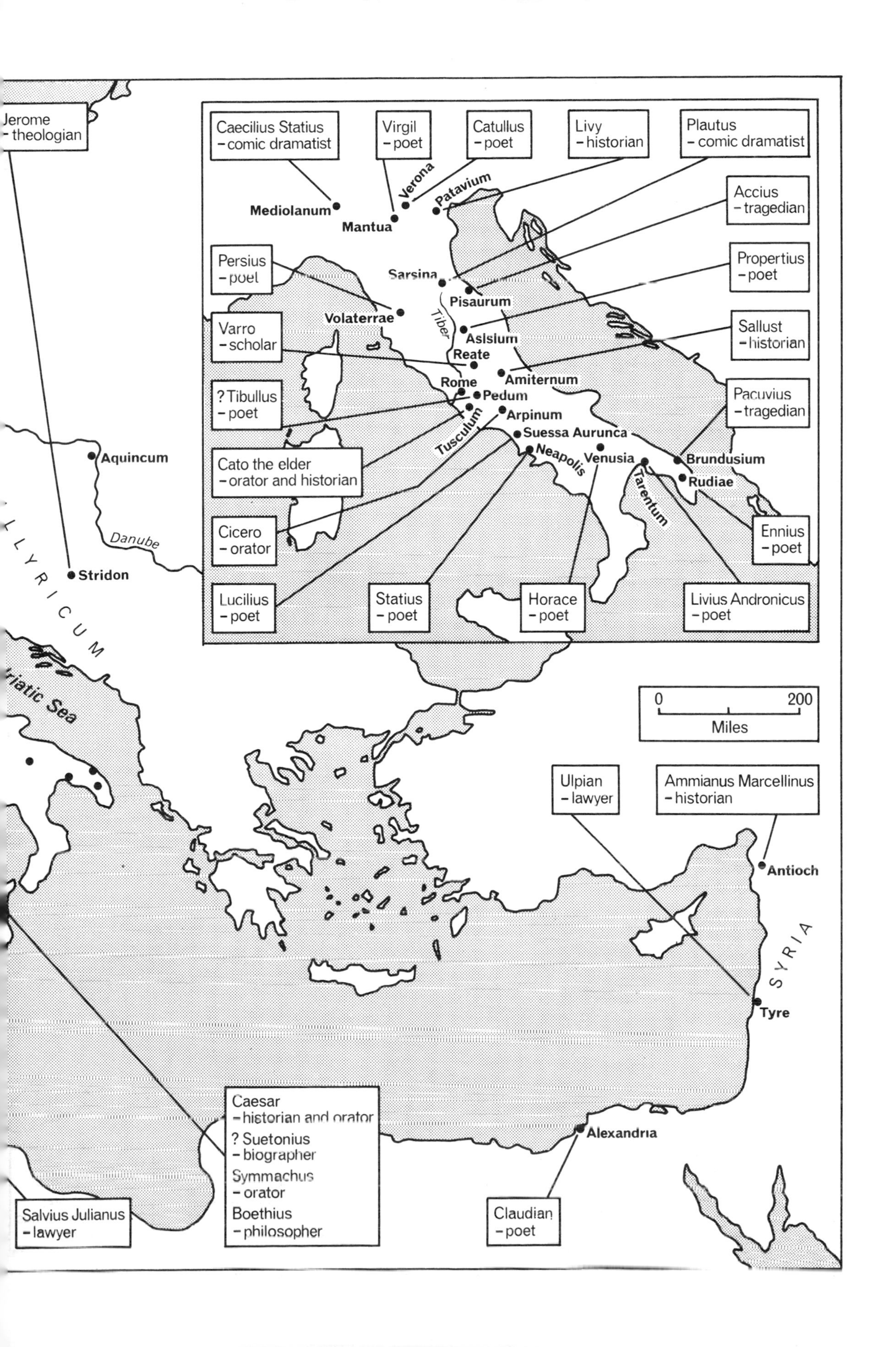
Jerome
- theologian
Caecilius Statius
- comic dramatist
Virgil
- poet
Catullus
- poet
Livy
- historian
Plautus
- comic dramatist
Accius
- tragedian
Persius
- poet
Propertius
- poet
Varro
- scholar
Sallust
- historian
? Tibullus
- poet
Pacuvius
- tragedian
Cato the elder
- orator and historian
Cicero
- orator
Ennius
- poet
Lucilius
- poet
Statius
- poet
Horace
- poet
Livius Andronicus
- poet
Mediolanum
Mantua
Verona
Patavium
Sarsina
Pisaurum
Volaterrae
Tiber
Asisium
Reate
Rome
Pedum
Amiternum
Arpinum
Tusculum
Suessa Aurunca
Neapolis
Venusia
Tarentum
Brundusium
Rudiae
Aquincum
Danube
Stridon
ILLYRICUM
Adriatic Sea
0
200
Miles
Ulpian
- lawyer
Ammianus Marcellinus
- historian
Antioch
SYRIA
Tyre
Caesar
- historian and orator
? Suetonius
- biographer
Symmachus
- orator
Boethius
- philosopher
Alexandria
Salvius Julianus
- lawyer
Claudian
- poet

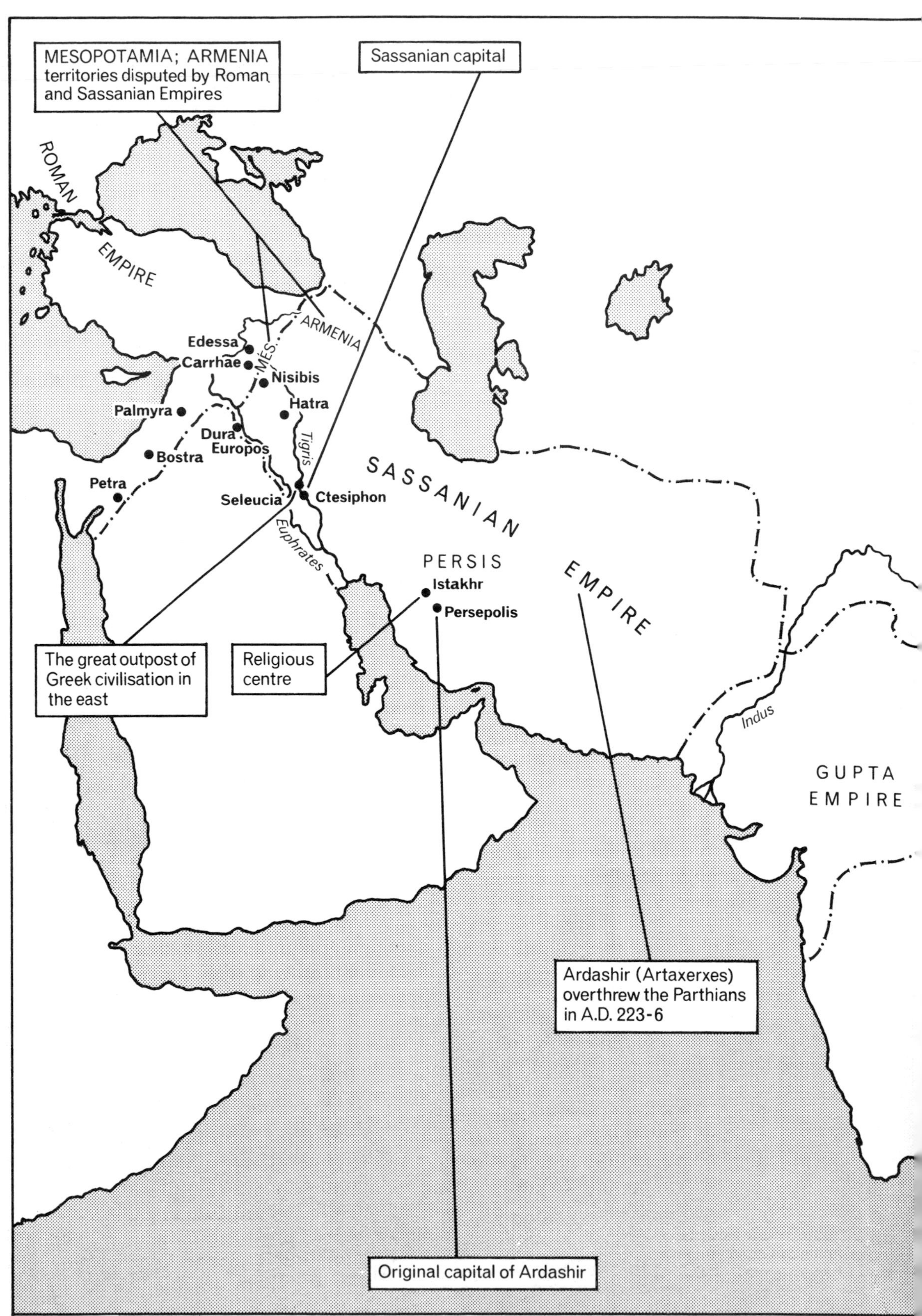
MESOPOTAMIA; ARMENIA
territories disputed by Roman and Sassanian Empires
Sassanian capital
ROMAN
EMPIRE
ARMENIA
MES.
Edessa
Carrhae
Nisibis
Hatra
Palmyra
Dura Europos
Bostra
Petra
Tigris
Seleucia
Ctesiphon
SASSANIAN
EMPIRE
Euphrates
PERSIS
Istakhr
Persepolis
Indus
GUPTA EMPIRE
The great outpost of Greek civilisation in the east
Religious centre
Ardashir (Artaxerxes) overthrew the Parthians in A.D. 223-6
Original capital of Ardashir

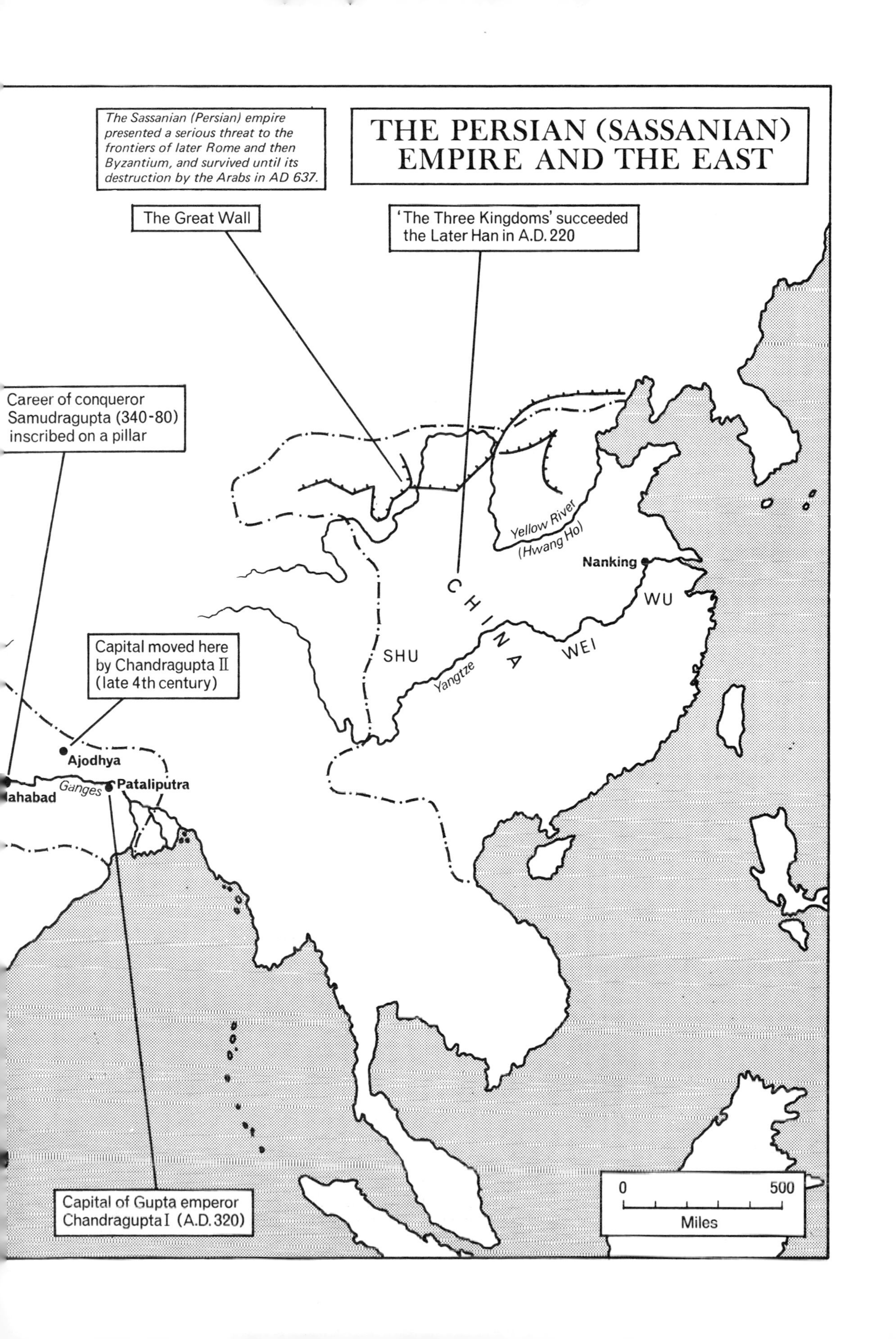

THE PERSIAN (SASSANIAN) EMPIRE AND THE EAST
The Sassanian (Persian) empire presented a serious threat to the frontiers of later Rome and then Byzantium, and survived until its destruction by the Arabs in AD 637.
The Great Wall
'The Three Kingdoms' succeeded the Later Han in A.D. 220
Career of conqueror Samudragupta (340-80) inscribed on a pillar
Capital moved here by Chandragupta II (late 4th century)
Capital of Gupta emperor Chandragupta I (A.D. 320)
Yellow River (Hwang Ho)
Nanking
CHINA
WU
WEI
SHU
Yangtze
Ajodhya
Pataliputra
Ganges
lahabad
0
500
Miles

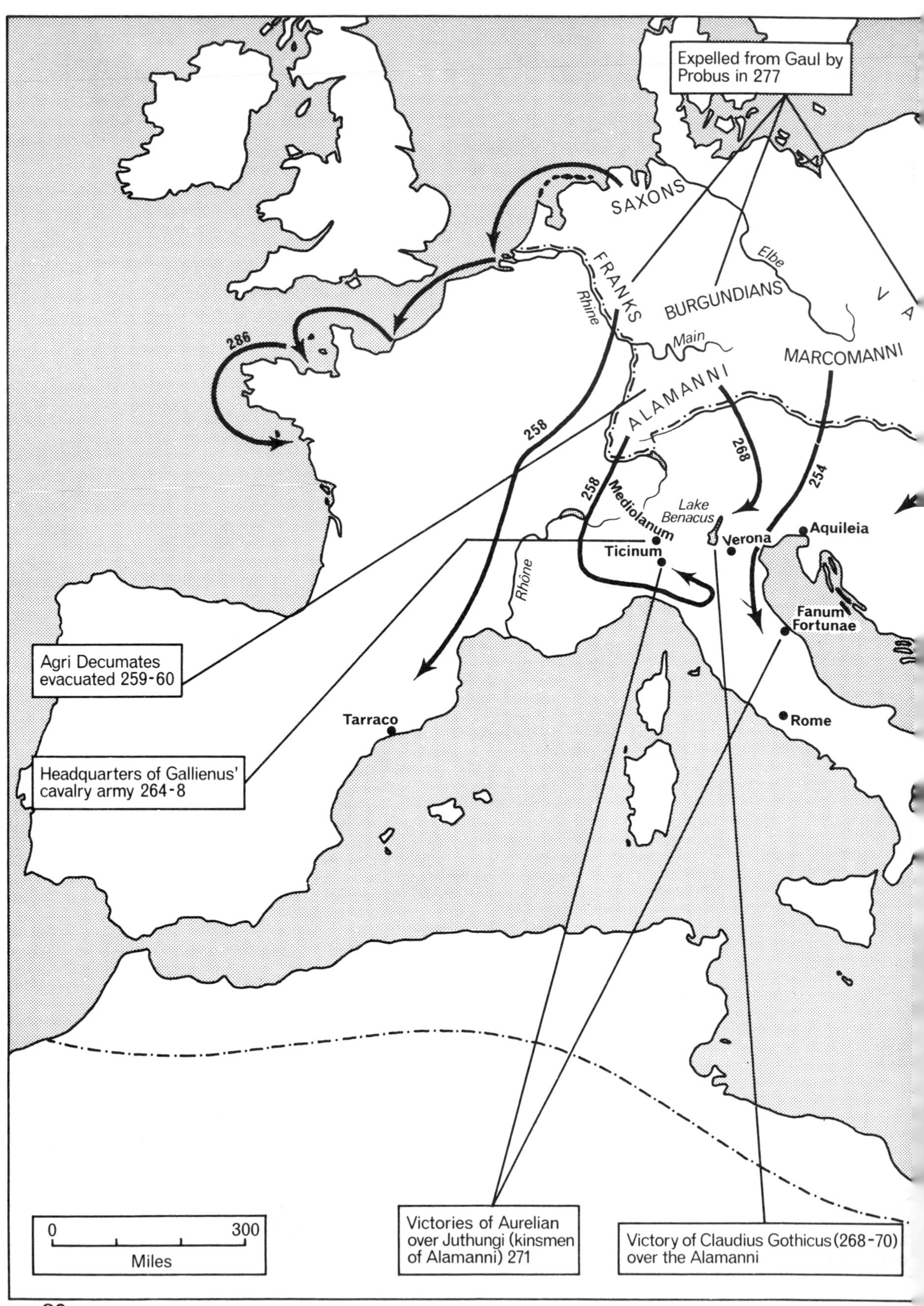

Expelled from Gaul by Probus in 277
SAXONS
FRANKS
BURGUNDIANS
Elbe
Rhine
Main
V A
MARCOMANNI
ALAMANNI
286
258
268
254
258
Mediolanum
Lake Benacus
Verona
Aquileia
Ticinum
Rhône
Fanum Fortunae
Agri Decumates evacuated 259-60
Tarraco
Rome
Headquarters of Gallienus' cavalry army 264-8
0
300
Miles
Victories of Aurelian over Juthungi (kinsmen of Alamanni) 271
Victory of Claudius Gothicus (268-70) over the Alamanni

GERMAN INVASIONS IN THE THIRD CENTURY A.D.

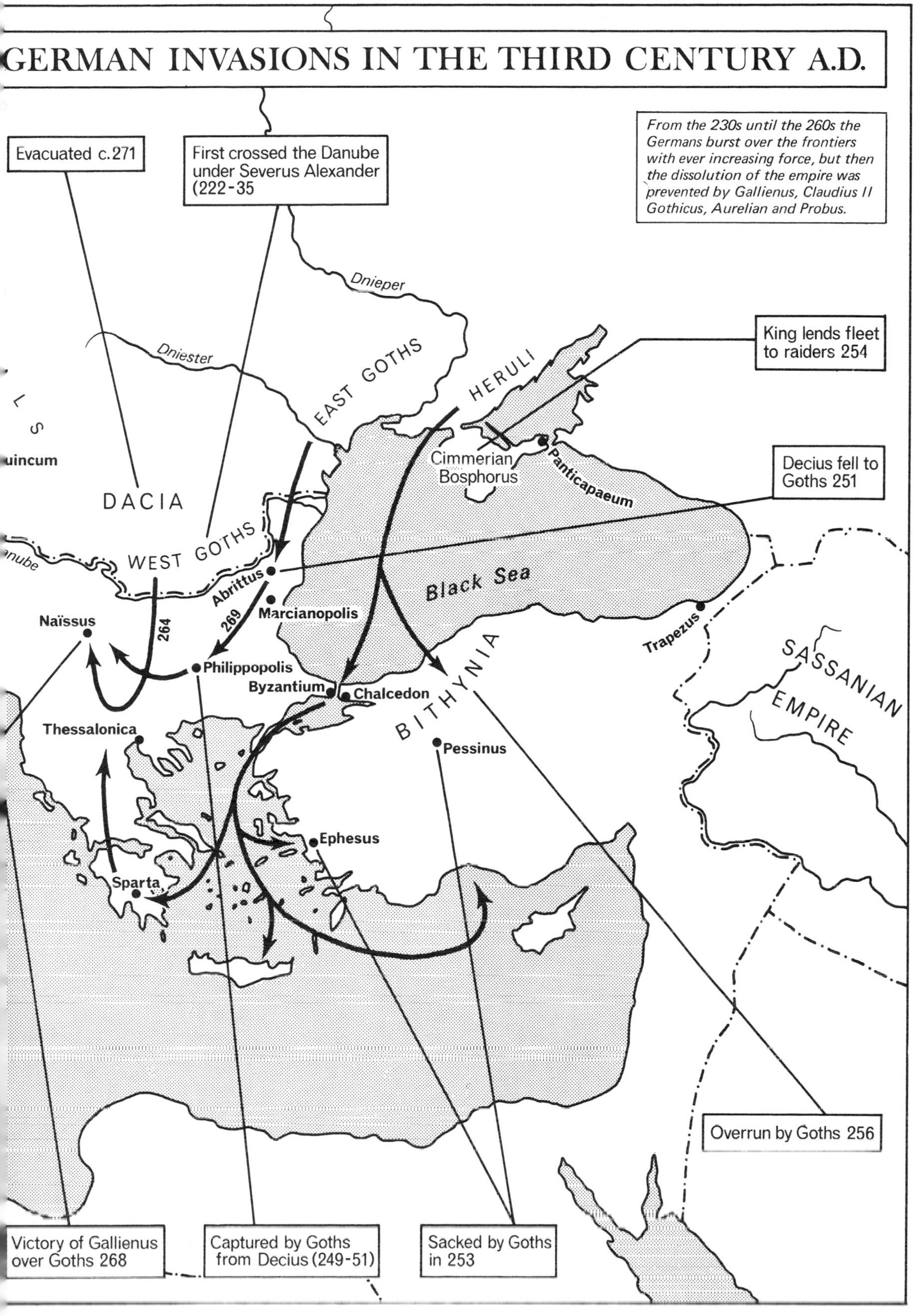

From the 230s until the 260s the Germans burst over the frontiers with ever increasing force, but then the dissolution of the empire was prevented by Gallienus, Claudius II Gothicus, Aurelian and Probus.

Evacuated c.271

First crossed the Danube under Severus Alexander (222-35

King lends fleet to raiders 254

Decius fell to Goths 251

Overrun by Goths 256

Victory of Gallienus over Goths 268

Captured by Goths from Decius (249-51)

Sacked by Goths in 253

Dnieper

Dniester

EAST GOTHS

HERULI

Cimmerian Bosphorus

Panticapaeum

uincum

DACIA

WEST GOTHS

nube

Abrittus

Marcianopolis

Black Sea

Naïssus

264

269

Philippopolis

Byzantium

Chalcedon

Trapezus

SASSANIAN EMPIRE

BITHYNIA

Pessinus

Thessalonica

Ephesus

Sparta

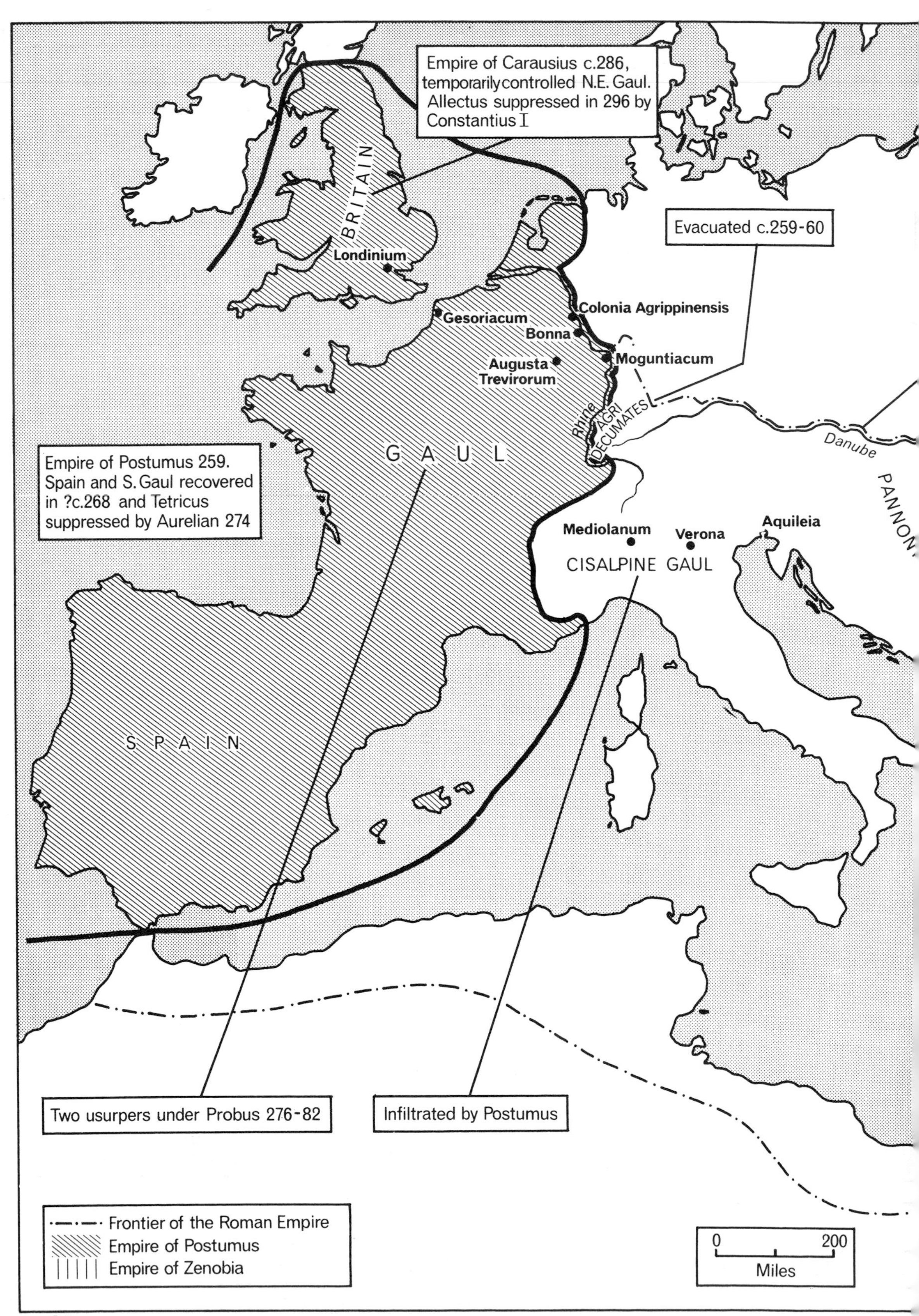
Empire of Carausius c.286, temporarily controlled N.E. Gaul. Allectus suppressed in 296 by Constantius I
BRITAIN
Londinium
Evacuated c.259-60
Colonia Agrippinensis
Gesoriacum
Bonna
Moguntiacum
Augusta Trevirorum
Rhine
AGRI DECUMATES
Danube
PANNON
Empire of Postumus 259. Spain and S. Gaul recovered in ?c.268 and Tetricus suppressed by Aurelian 274
GAUL
Mediolanum
Verona
Aquileia
CISALPINE GAUL
SPAIN
Two usurpers under Probus 276-82
Infiltrated by Postumus
Frontier of the Roman Empire
Empire of Postumus
Empire of Zenobia
0
200
Miles

THE BREAKDOWN AND RECOVERY OF THE ROMAN EMPIRE IN THE LATER THIRD CENTURY A.D.

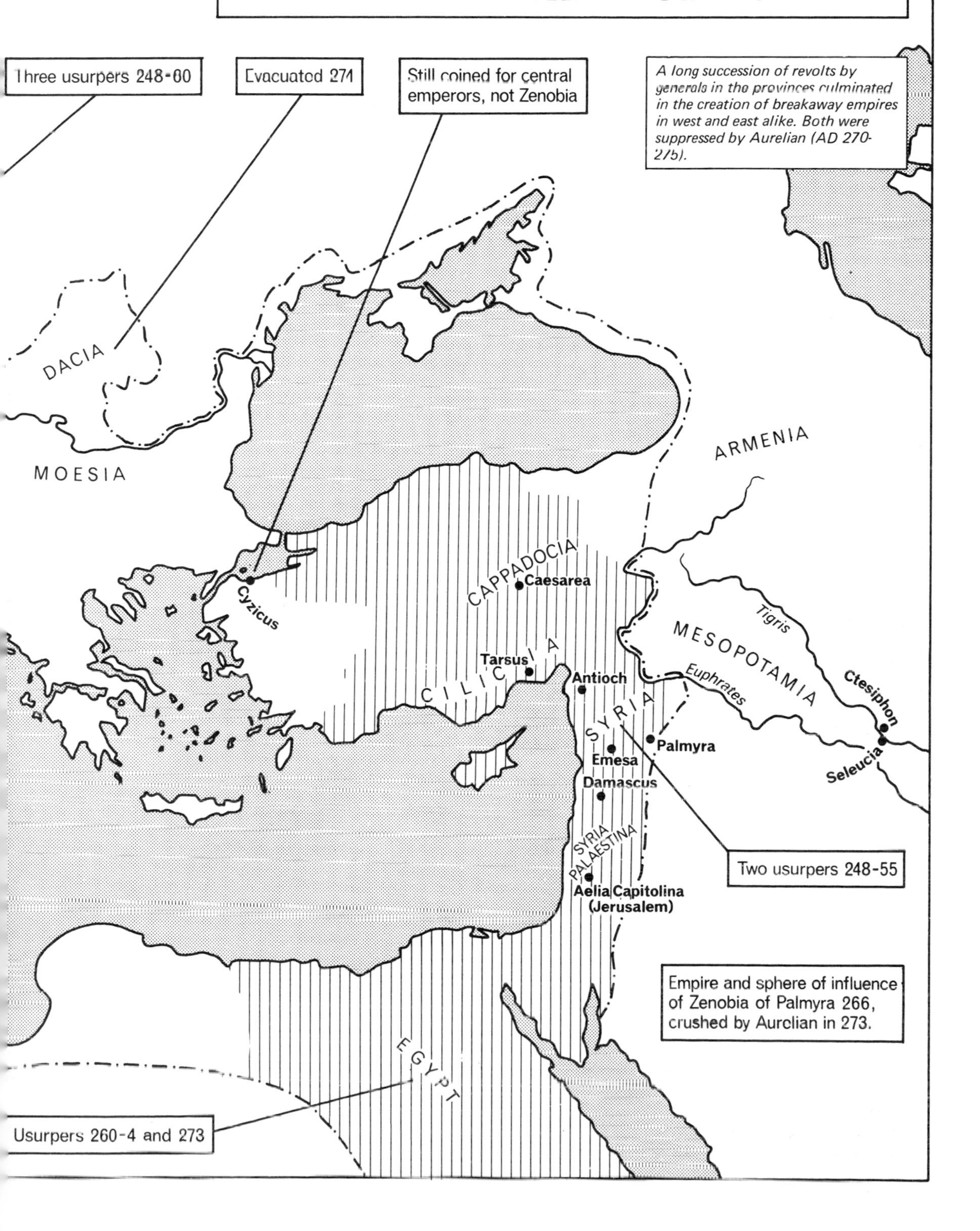

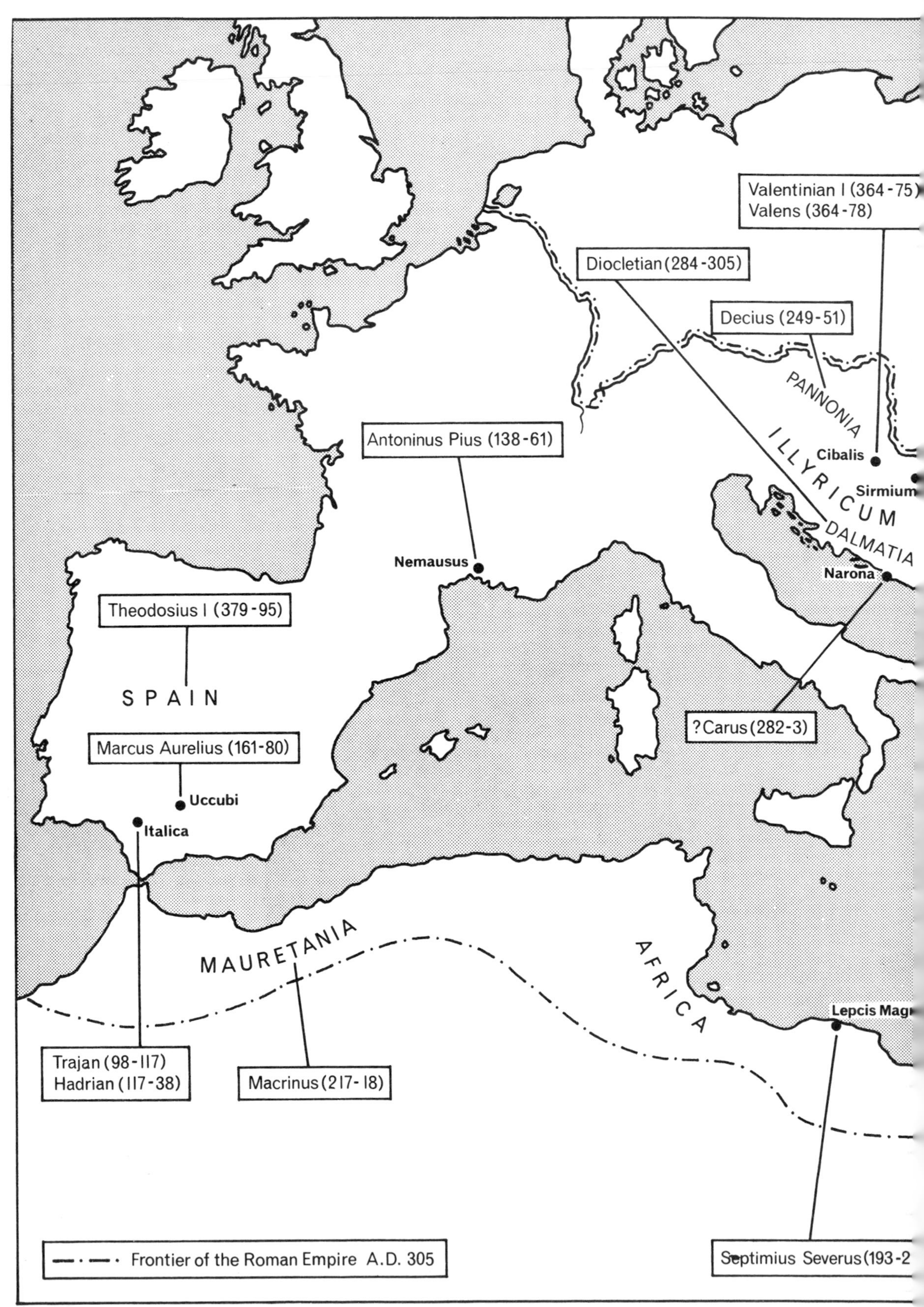
Valentinian I (364-75)
Valens (364-78)
Diocletian (284-305)
Decius (249-51)
PANNONIA
ILLYRICUM
Cibalis
Sirmium
Antoninus Pius (138-61)
DALMATIA
Narona
Nemausus
Theodosius I (379-95)
SPAIN
Marcus Aurelius (161-80)
Uccubi
Italica
?Carus (282-3)
MAURETANIA
AFRICA
Lepcis Mag
Trajan (98-117)
Hadrian (117-38)
Macrinus (217-18)
Frontier of the Roman Empire A.D. 305
Septimius Severus (193-2

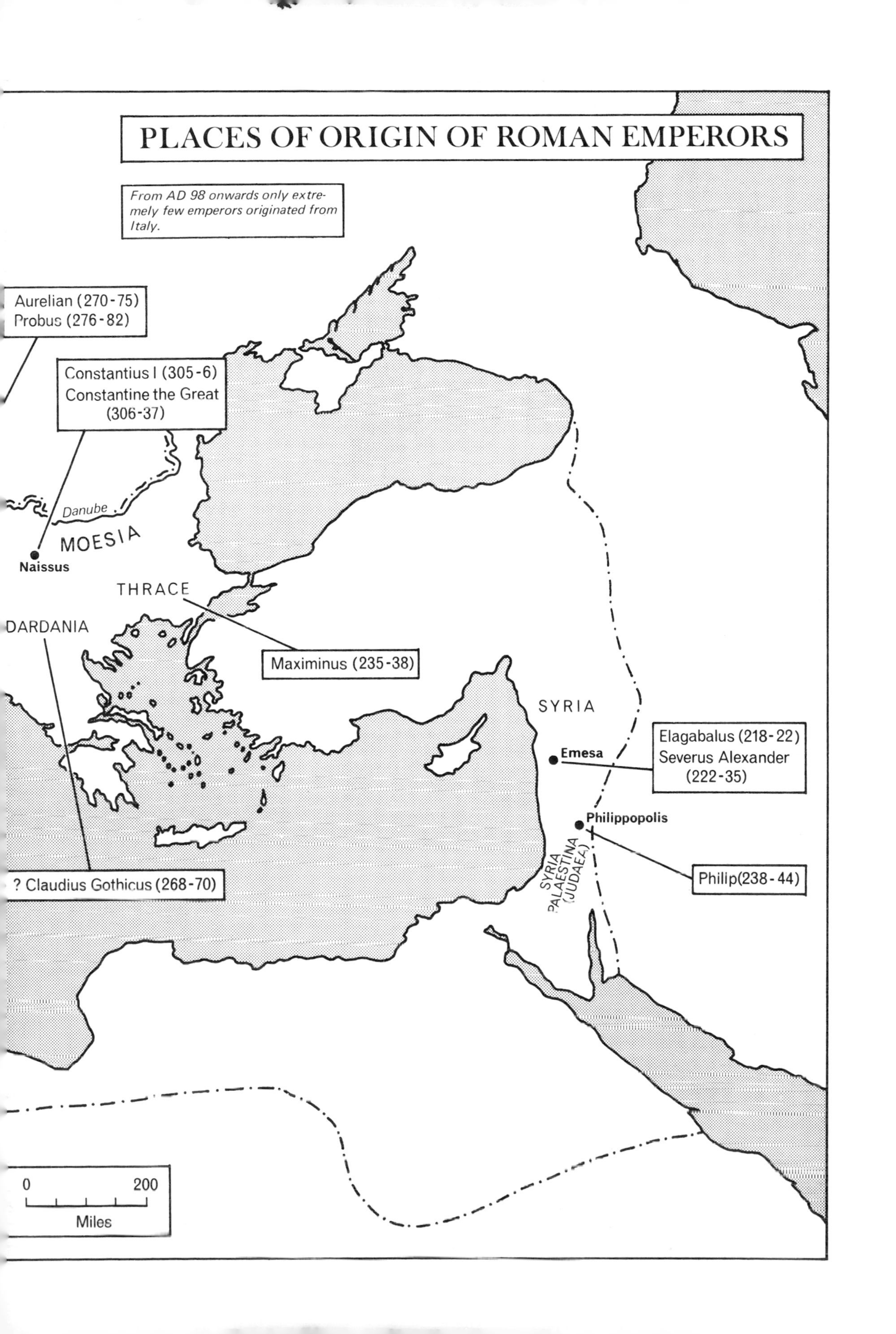

PLACES OF ORIGIN OF ROMAN EMPERORS
From AD 98 onwards only extremely few emperors originated from Italy.
Aurelian (270-75)
Probus (276-82)
Constantius I (305-6)
Constantine the Great (306-37)
Danube
MOESIA
Naissus
THRACE
DARDANIA
Maximinus (235-38)
SYRIA
Emesa
Elagabalus (218-22)
Severus Alexander (222-35)
Philippopolis
Philip(238-44)
SYRIA PALAESTINA (JUDAEA)
? Claudius Gothicus (268-70)
0
200
Miles

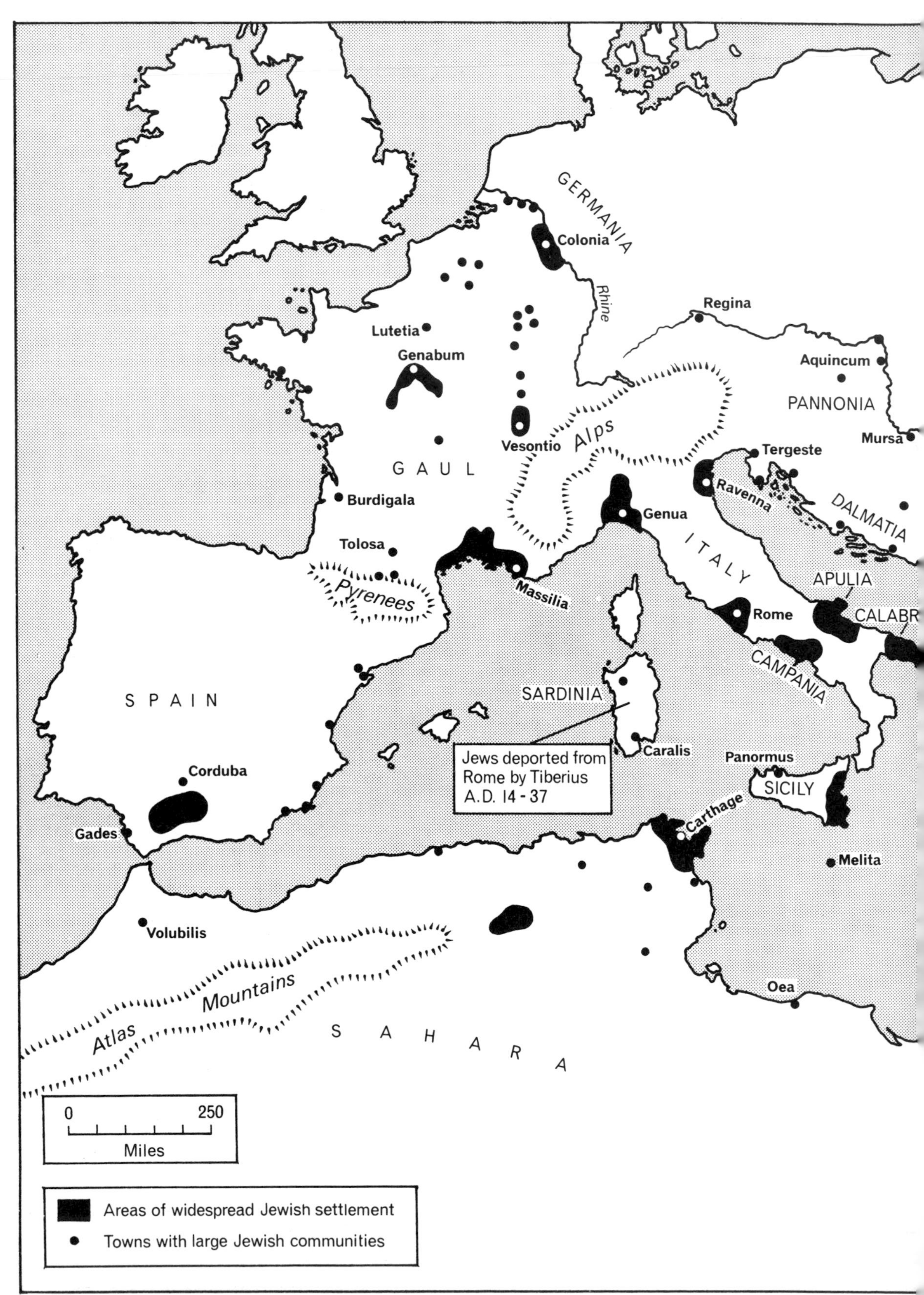
GERMANIA
Colonia
Rhine
Regina
Lutetia
Genabum
Aquincum
PANNONIA
Alps
Vesontio
Mursa
Tergeste
GAUL
Ravenna
DALMATIA
Burdigala
Genua
ITALY
Tolosa
Pyrenees
Massilia
APULIA
Rome
CALABR
CAMPANIA
SPAIN
SARDINIA
Caralis
Jews deported from
Rome by Tiberius
A.D. 14 - 37
Panormus
Corduba
SICILY
Carthage
Gades
Melita
Volubilis
Mountains
Oea
Atlas
SAHARA
0
250
Miles
Areas of widespread Jewish settlement
Towns with large Jewish communities

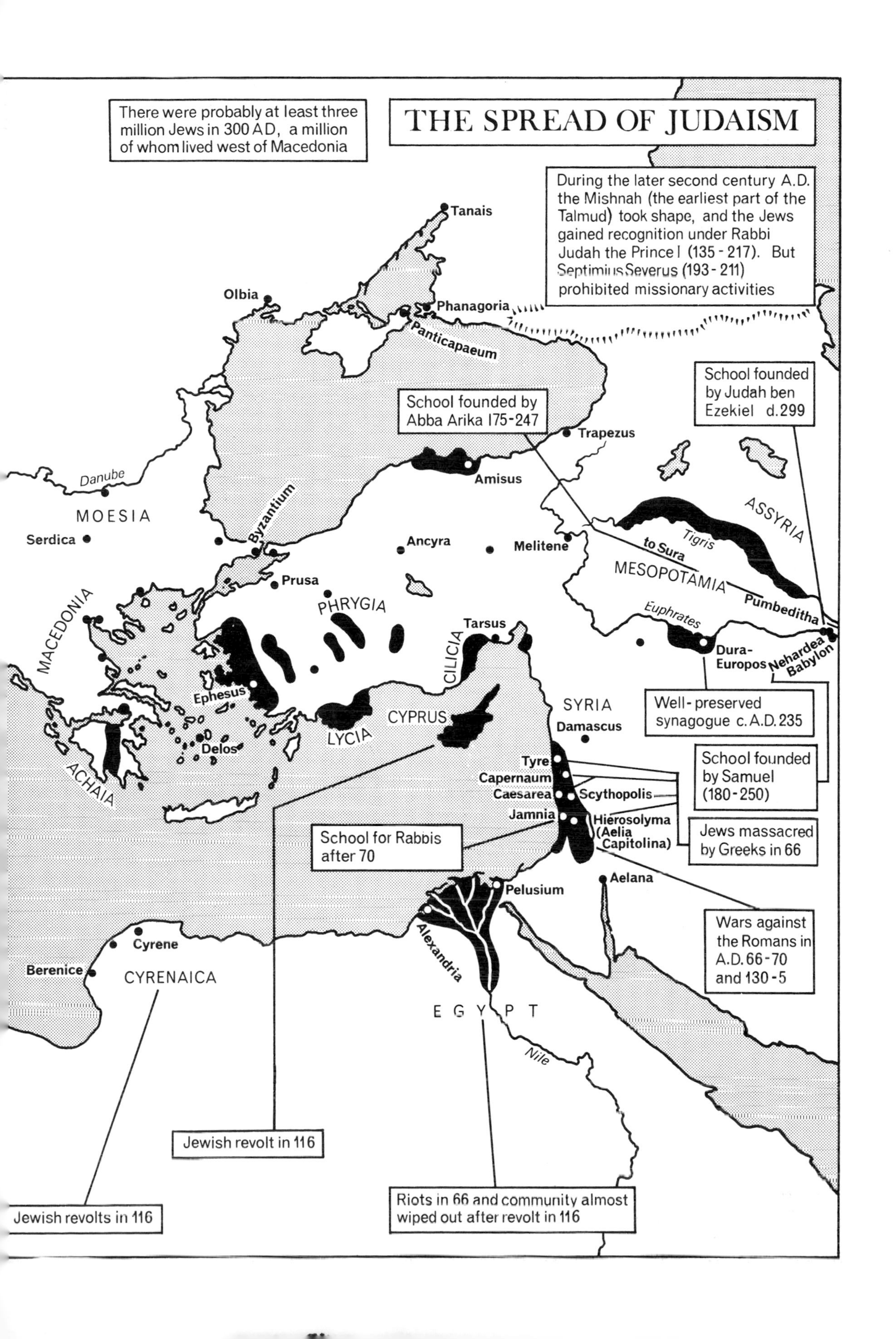

THE SPREAD OF JUDAISM
There were probably at least three million Jews in 300 AD, a million of whom lived west of Macedonia
During the later second century A.D. the Mishnah (the earliest part of the Talmud) took shape, and the Jews gained recognition under Rabbi Judah the Prince I (135 - 217). But Septimius Severus (193 - 211) prohibited missionary activities
Tanais
Olbia
Phanagoria
Panticapaeum
School founded by Abba Arika 175-247
School founded by Judah ben Ezekiel d.299
Trapezus
Danube
Amisus
MOESIA
Byzantium
Serdica
Ancyra
Melitene
ASSYRIA
Tigris
to Sura
MESOPOTAMIA
Prusa
PHRYGIA
MACEDONIA
Pumbeditha
Euphrates
Tarsus
CILICIA
Dura-Europos
Nehardea
Babylon
Ephesus
SYRIA
Well-preserved synagogue c. A.D. 235
CYPRUS
LYCIA
Damascus
Delos
ACHAIA
Tyre
Capernaum
Caesarea
Scythopolis
Jamnia
Hierosolyma (Aelia Capitolina)
School founded by Samuel (180-250)
Jews massacred by Greeks in 66
School for Rabbis after 70
Aelana
Pelusium
Wars against the Romans in A.D. 66-70 and 130-5
Cyrene
Berenice
CYRENAICA
Alexandria
EGYPT
Nile
Jewish revolt in 116
Jewish revolts in 116
Riots in 66 and community almost wiped out after revolt in 116

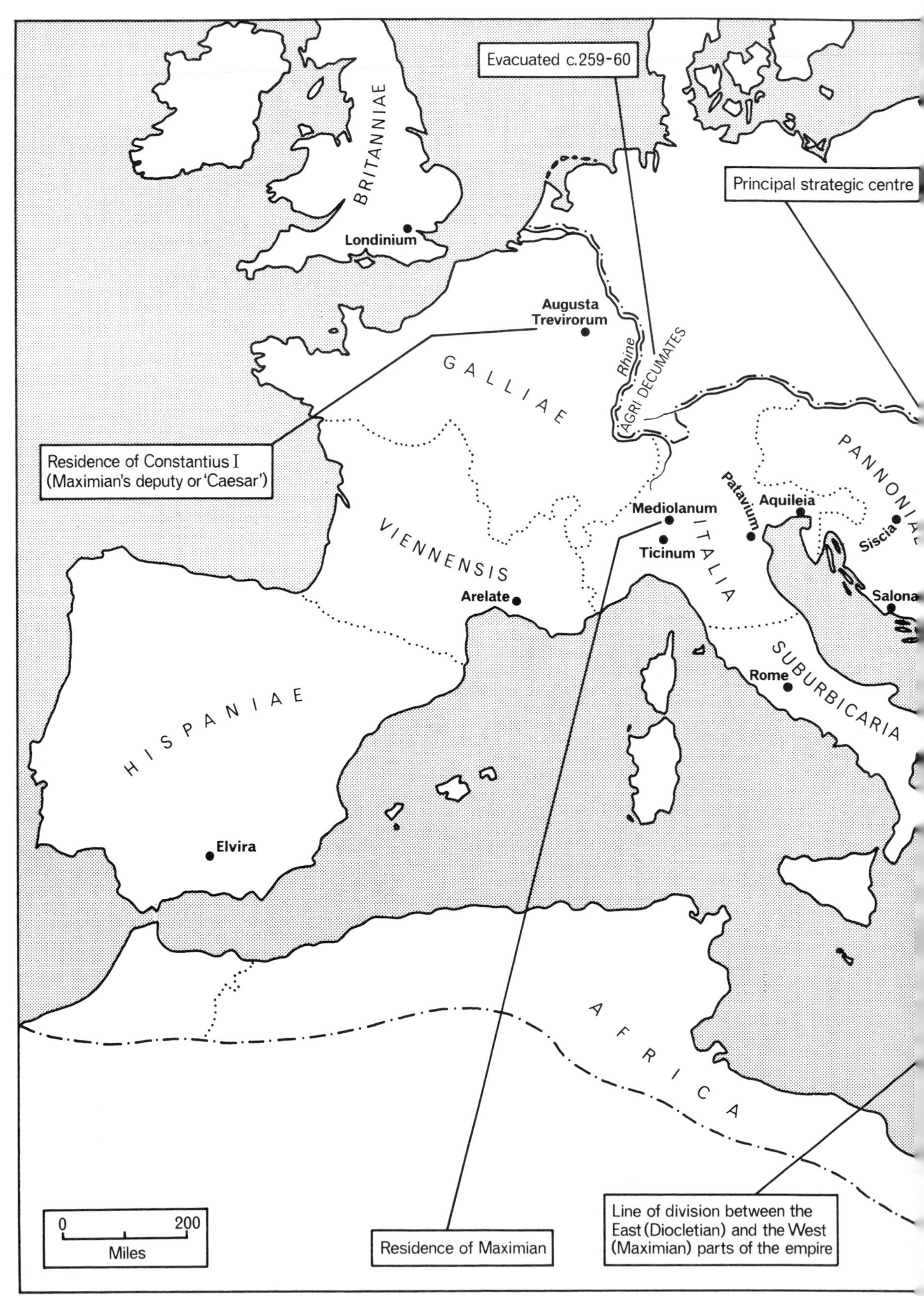

Evacuated c.259-60
Principal strategic centre
BRITANNIAE
Londinium
Augusta Trevirorum
Rhine
AGRI DECUMATES
GALLIAE
Residence of Constantius I (Maximian's deputy or 'Caesar')
PANNONIAE
Mediolanum
Patavium
Aquileia
Siscia
VIENNENSIS
Ticinum
ITALIA
Arelate
Salona
SUBURBICARIA
Rome
HISPANIAE
Elvira
AFRICA
0 200 Miles
Residence of Maximian
Line of division between the East (Diocletian) and the West (Maximian) parts of the empire

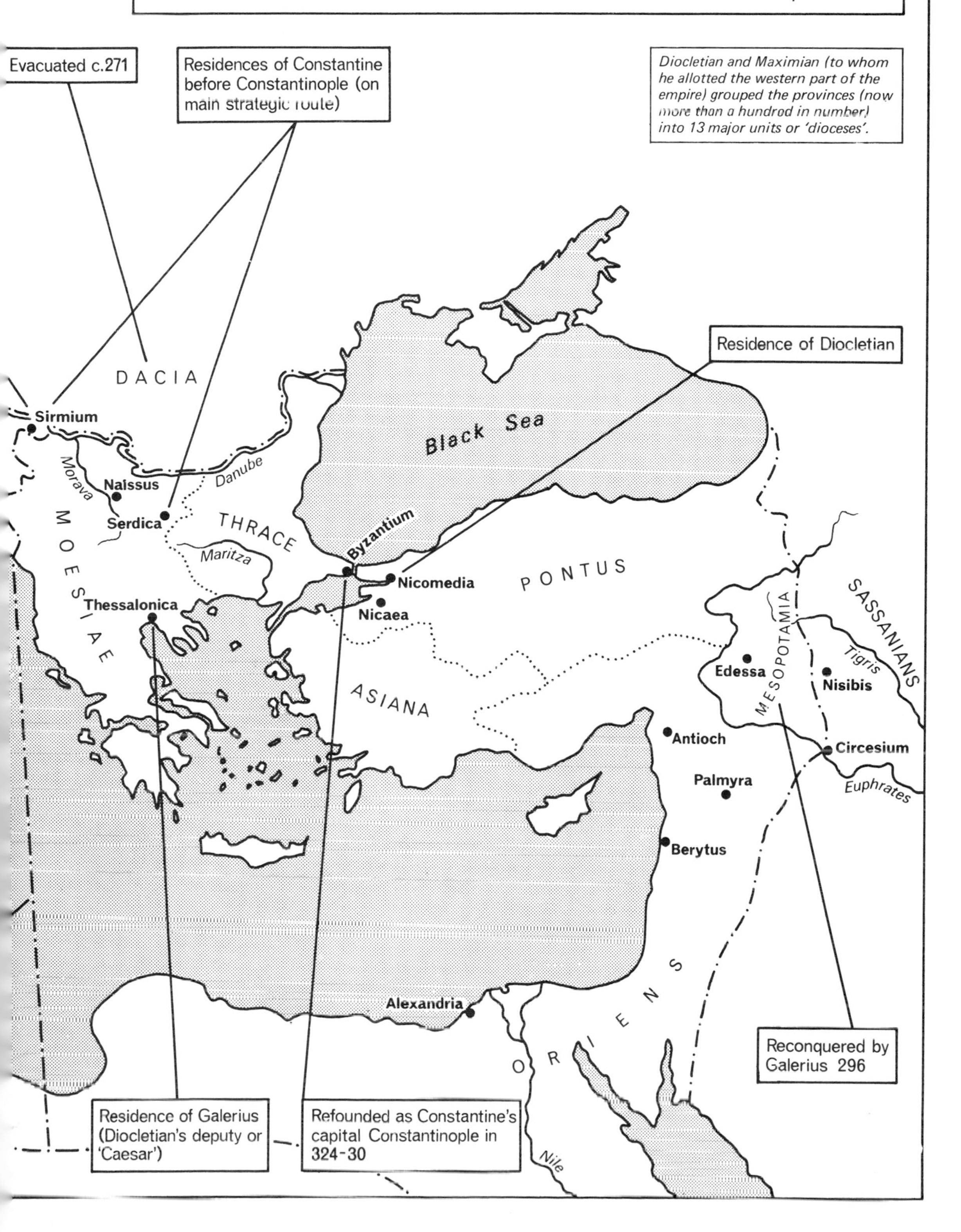
THE ROMAN EMPIRE UNDER DIOCLETIAN AND MAXIMIAN A.D. 284/6-305
Evacuated c.271
Residences of Constantine before Constantinople (on main strategic route)
Diocletian and Maximian (to whom he allotted the western part of the empire) grouped the provinces (now more than a hundred in number) into 13 major units or 'dioceses'.
Residence of Diocletian
DACIA
Sirmium
Morava
Danube
Naissus
Serdica
THRACE
Maritza
MOESIAE
Black Sea
Byzantium
Nicomedia
Nicaea
PONTUS
Thessalonica
ASIANA
SASSANIANS
MESOPOTAMIA
Tigris
Edessa
Nisibis
Antioch
Circesium
Palmyra
Euphrates
Berytus
ORIENS
Alexandria
Nile
Reconquered by Galerius 296
Residence of Galerius (Diocletian's deputy or 'Caesar')
Refounded as Constantine's capital Constantinople in 324-30

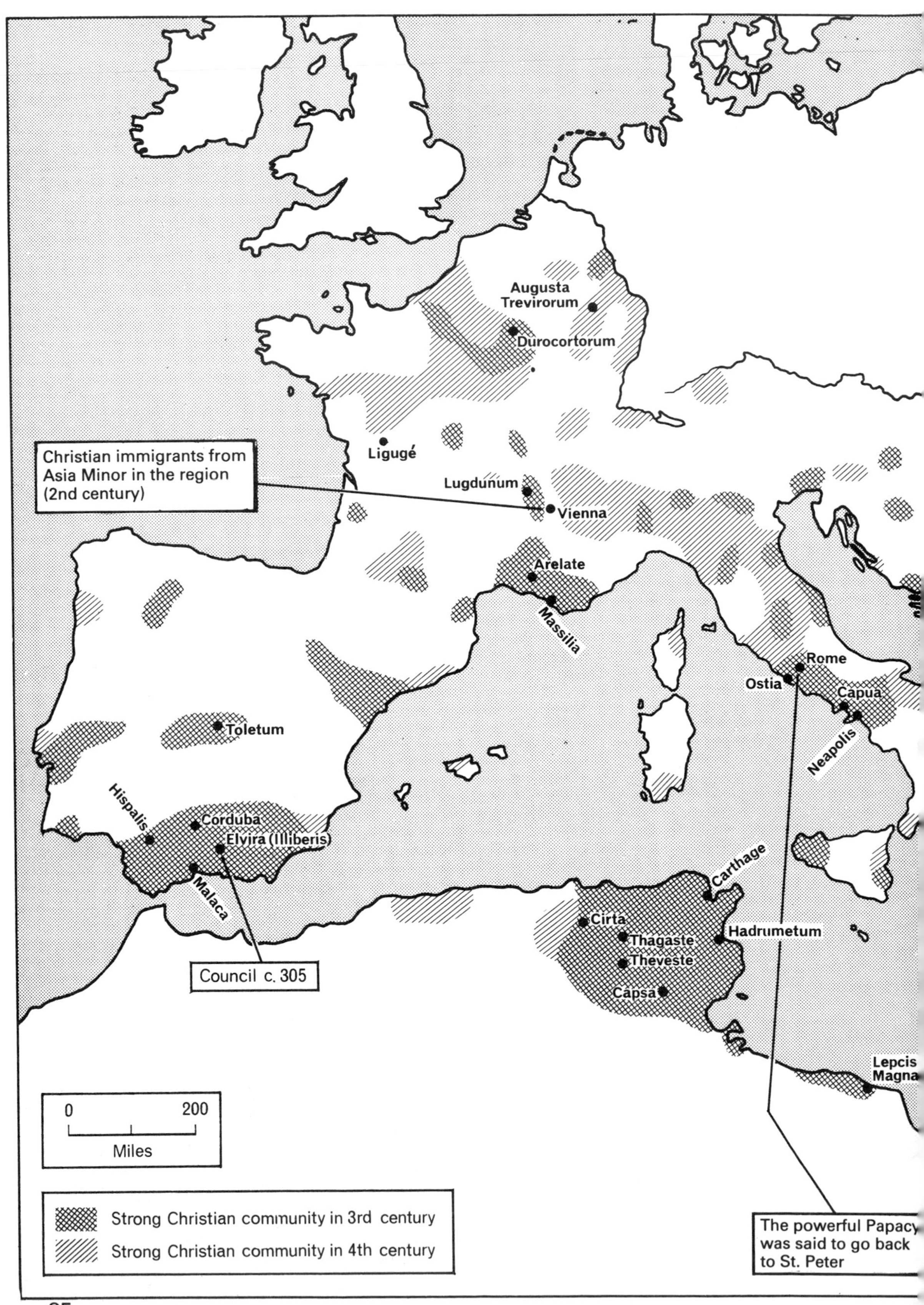

Augusta
Trevirorum
Durocortorum
Liguge
Christian immigrants from
Asia Minor in the region
(2nd century)
Lugdunum
Vienna
Arelate
Massilia
Rome
Ostia
Capua
Neapolis
Toletum
Hispalis
Corduba
Elvira (Illiberis)
Malaca
Council c. 305
Carthage
Cirta
Thagaste
Theveste
Hadrumetum
Capsa
Lepcis
Magna
0
200
Miles
Strong Christian community in 3rd century
Strong Christian community in 4th century
The powerful Papacy
was said to go back
to St. Peter

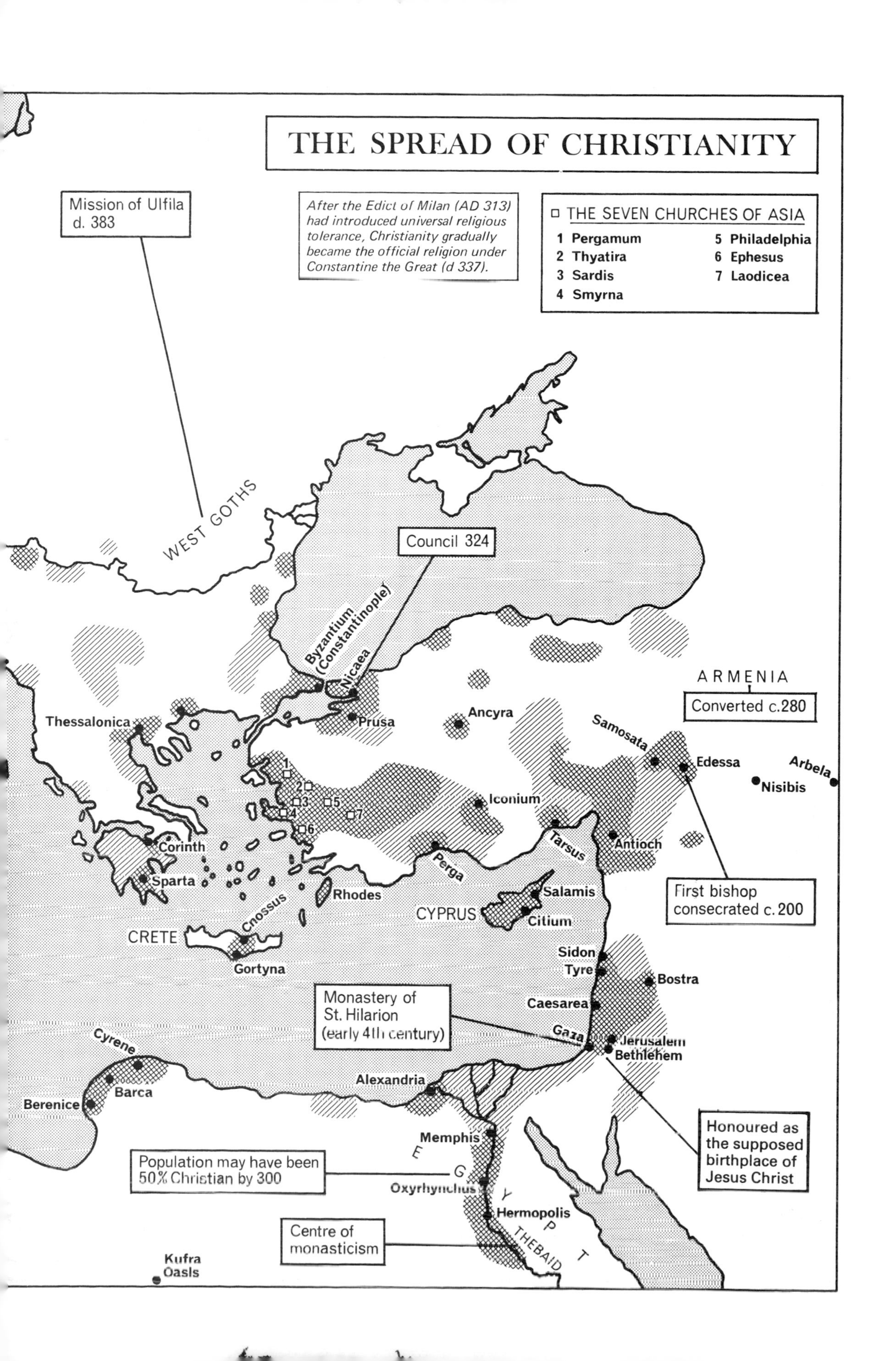
THE SPREAD OF CHRISTIANITY
Mission of Ulfila
d. 383
After the Edict of Milan (AD 313) had introduced universal religious tolerance, Christianity gradually became the official religion under Constantine the Great (d 337).
THE SEVEN CHURCHES OF ASIA
1 Pergamum
2 Thyatira
3 Sardis
4 Smyrna
5 Philadelphia
6 Ephesus
7 Laodicea
WEST GOTHS
Council 324
Byzantium (Constantinople)
Nicaea
Prusa
Ancyra
ARMENIA
Converted c.280
Samosata
Edessa
Arbela
Nisibis
Thessalonica
Iconium
Antioch
Tarsus
Perga
Corinth
Sparta
Rhodes
Cnossus
CRETE
Gortyna
CYPRUS
Salamis
Citium
First bishop consecrated c. 200
Sidon
Tyre
Bostra
Caesarea
Monastery of St. Hilarion (early 4th century)
Gaza
Jerusalem
Bethlehem
Cyrene
Barca
Berenice
Alexandria
Memphis
Honoured as the supposed birthplace of Jesus Christ
Population may have been 50% Christian by 300
EGYPT
Oxyrhynchus
Hermopolis
THEBAID
Centre of monasticism
Kufra Oasis

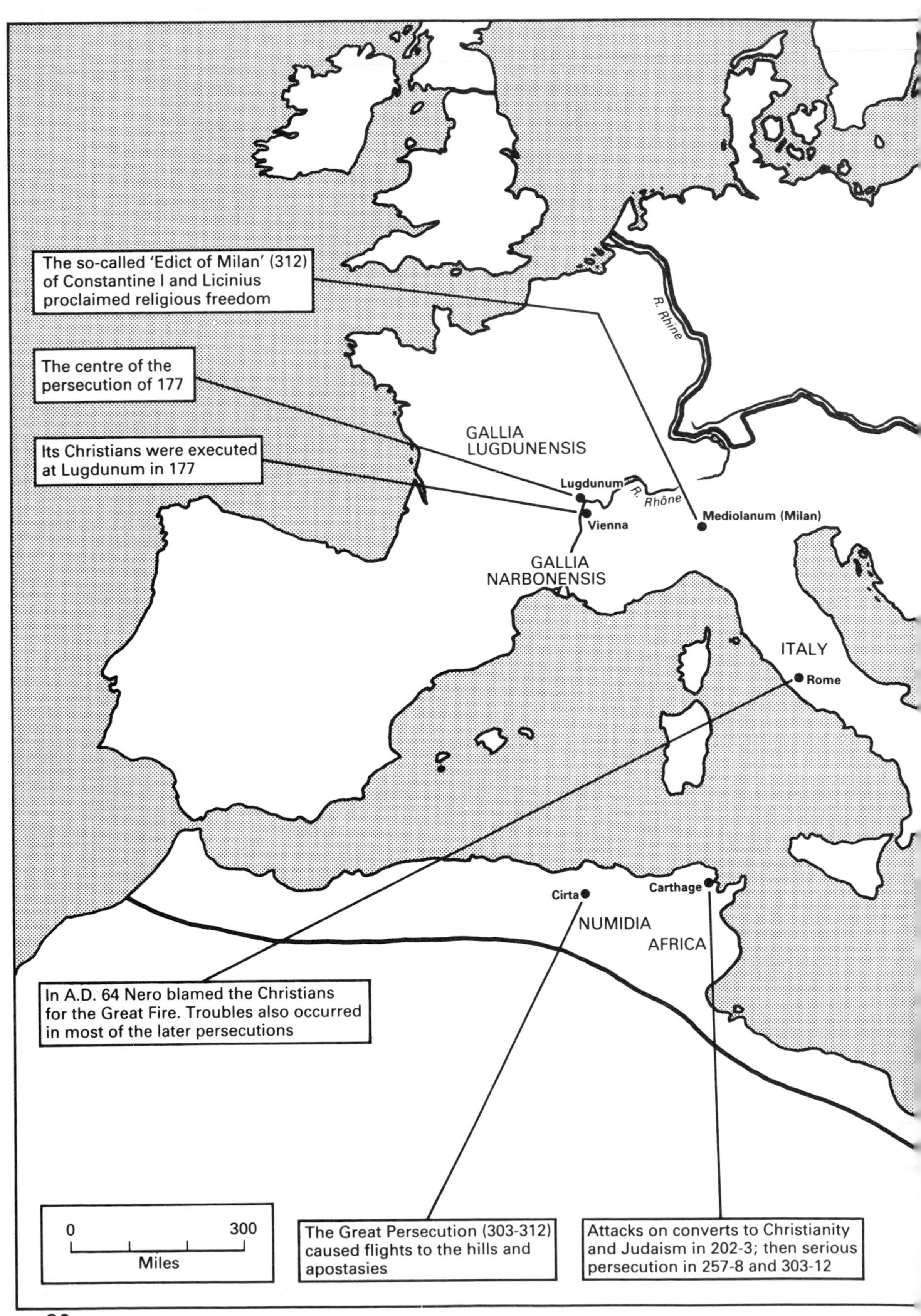
The so-called 'Edict of Milan' (312) of Constantine I and Licinius proclaimed religious freedom
The centre of the persecution of 177
Its Christians were executed at Lugdunum in 177
R. Rhine
GALLIA LUGDUNENSIS
Lugdunum
R. Rhône
Vienna
Mediolanum (Milan)
GALLIA NARBONENSIS
ITALY
Rome
Cirta
Carthage
NUMIDIA
AFRICA
In A.D. 64 Nero blamed the Christians for the Great Fire. Troubles also occurred in most of the later persecutions
0
300
Miles
The Great Persecution (303-312) caused flights to the hills and apostasies
Attacks on converts to Christianity and Judaism in 202-3; then serious persecution in 257-8 and 303-12

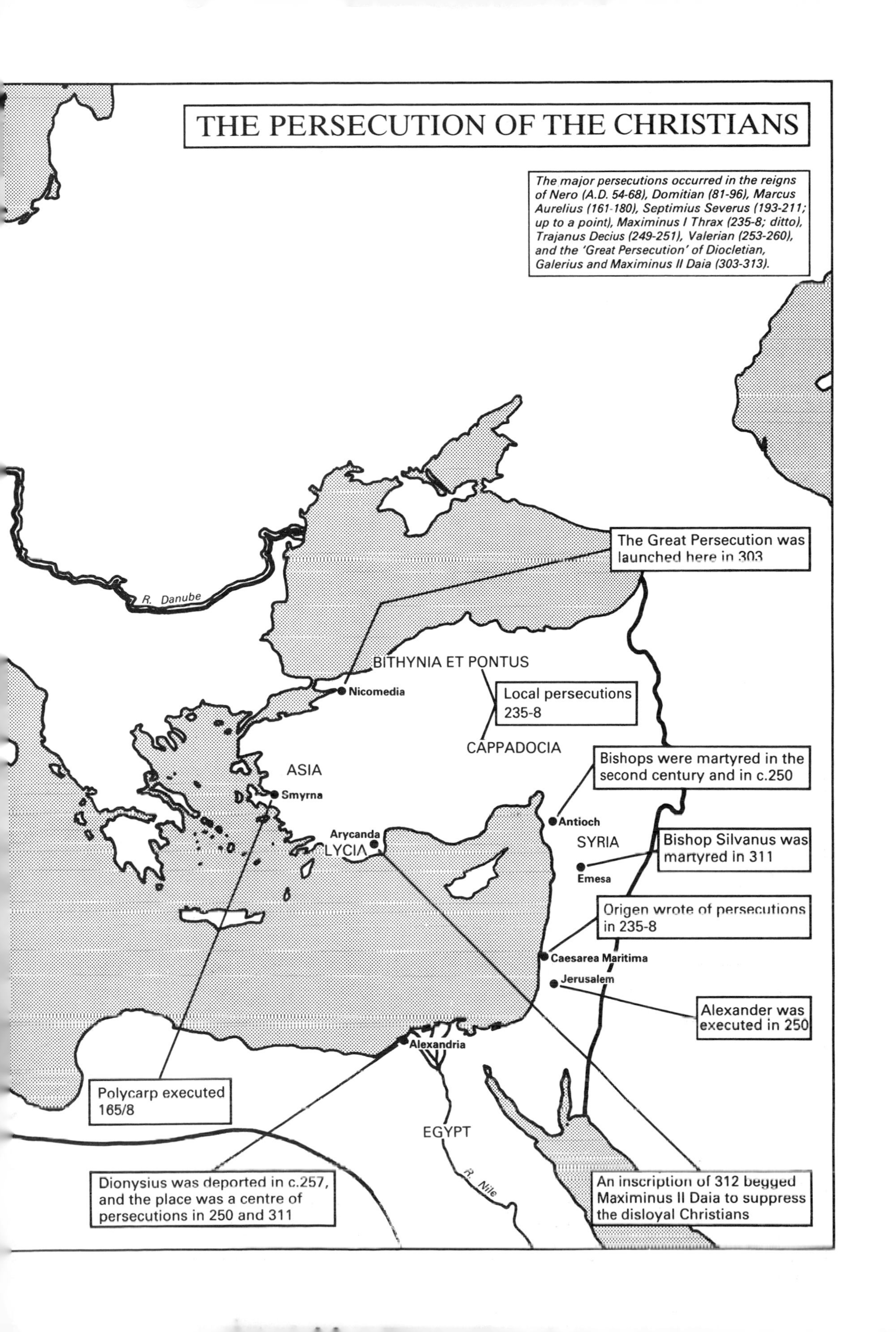
THE PERSECUTION OF THE CHRISTIANS
The major persecutions occurred in the reigns of Nero (A.D. 54-68), Domitian (81-96), Marcus Aurelius (161-180), Septimius Severus (193-211; up to a point), Maximinus I Thrax (235-8; ditto), Trajanus Decius (249-251), Valerian (253-260), and the 'Great Persecution' of Diocletian, Galerius and Maximinus II Daia (303-313).
R. Danube
The Great Persecution was launched here in 303
BITHYNIA ET PONTUS
Nicomedia
Local persecutions 235-8
CAPPADOCIA
ASIA
Smyrna
Bishops were martyred in the second century and in c.250
Arycanda
LYCIA
Antioch
SYRIA
Bishop Silvanus was martyred in 311
Emesa
Origen wrote of persecutions in 235-8
Caesarea Maritima
Jerusalem
Alexander was executed in 250
Alexandria
Polycarp executed 165/8
EGYPT
R. Nile
Dionysius was deported in c.257, and the place was a centre of persecutions in 250 and 311
An inscription of 312 begged Maximinus II Daia to suppress the disloyal Christians

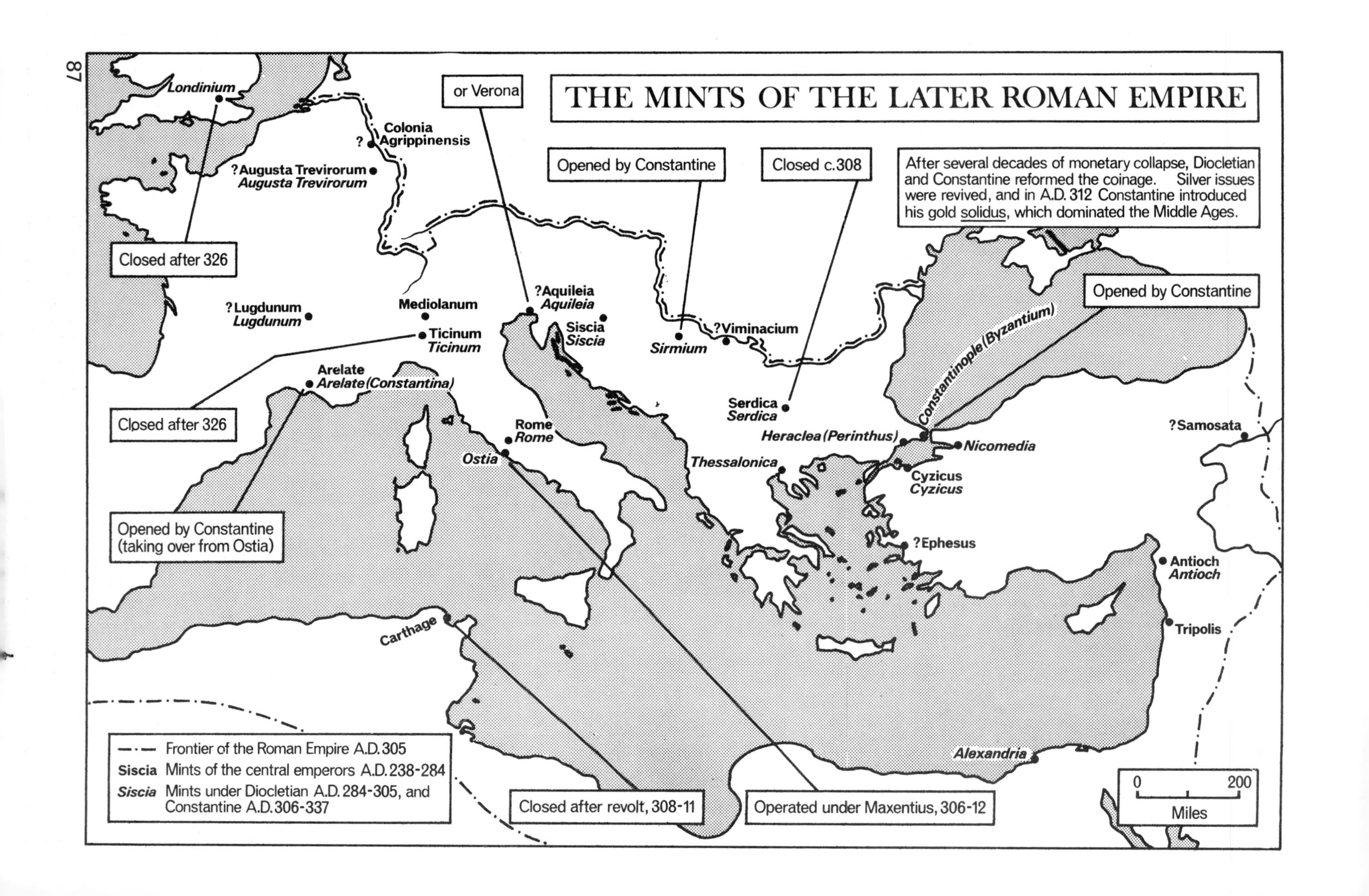
THE MINTS OF THE LATER ROMAN EMPIRE
After several decades of monetary collapse, Diocletian and Constantine reformed the coinage. Silver issues were revived, and in A.D. 312 Constantine introduced his gold solidus, which dominated the Middle Ages.
or Verona
Opened by Constantine
Closed c.308
Opened by Constantine
Closed after 326
Closed after 326
Opened by Constantine (taking over from Ostia)
Closed after revolt, 308-11
Operated under Maxentius, 306-12
Londinium
? Colonia Agrippinensis
?Augusta Trevirorum
Augusta Trevirorum
?Lugdunum
Lugdunum
Mediolanum
Ticinum
Ticinum
Arelate
Arelate (Constantina)
?Aquileia
Aquileia
Siscia
Siscia
Sirmium
?Viminacium
Serdica
Serdica
Rome
Rome
Ostia
Carthage
Thessalonica
Heraclea (Perinthus)
Constantinople (Byzantium)
Nicomedia
Cyzicus
Cyzicus
?Ephesus
?Samosata
Antioch
Antioch
Tripolis
Alexandria
0 200
Miles
Frontier of the Roman Empire A.D. 305
Siscia Mints of the central emperors A.D. 238-284
Siscia Mints under Diocletian A.D. 284-305, and Constantine A.D. 306-337

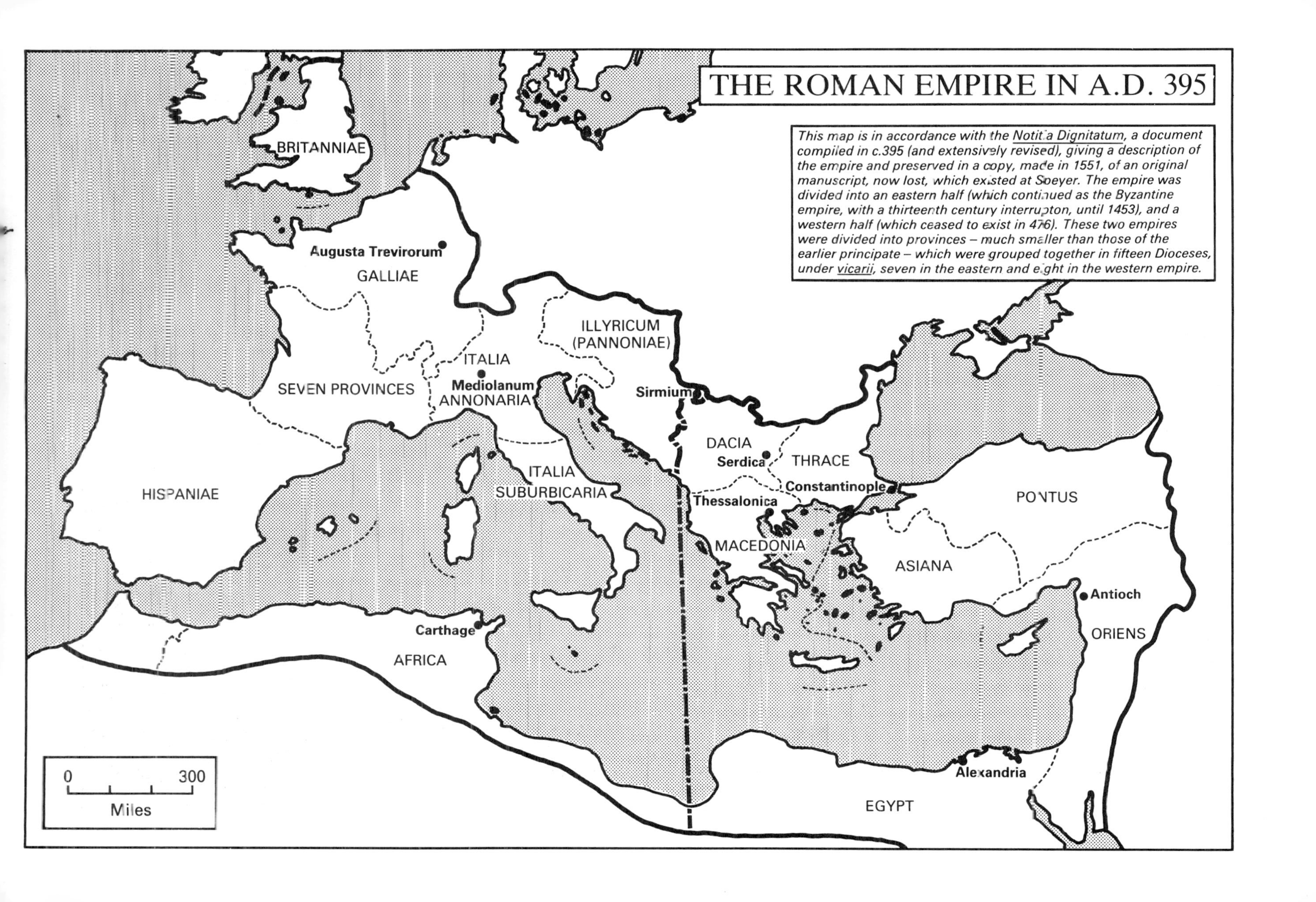
THE ROMAN EMPIRE IN A.D. 395
This map is in accordance with the Notitia Dignitatum, a document compiled in c.395 (and extensively revised), giving a description of the empire and preserved in a copy, made in 1551, of an original manuscript, now lost, which existed at Speyer. The empire was divided into an eastern half (which continued as the Byzantine empire, with a thirteenth century interruption, until 1453), and a western half (which ceased to exist in 476). These two empires were divided into provinces – much smaller than those of the earlier principate – which were grouped together in fifteen Dioceses, under vicarii, seven in the eastern and eight in the western empire.
BRITANNIAE
Augusta Trevirorum
GALLIAE
SEVEN PROVINCES
HISPANIAE
ITALIA
Mediolanum
ANNONARIA
ILLYRICUM
(PANNONIAE)
Sirmium
ITALIA
SUBURBICARIA
Carthage
AFRICA
DACIA
Serdica
THRACE
Constantinople
Thessalonica
MACEDONIA
PONTUS
ASIANA
Antioch
ORIENS
Alexandria
EGYPT
0
300
Miles

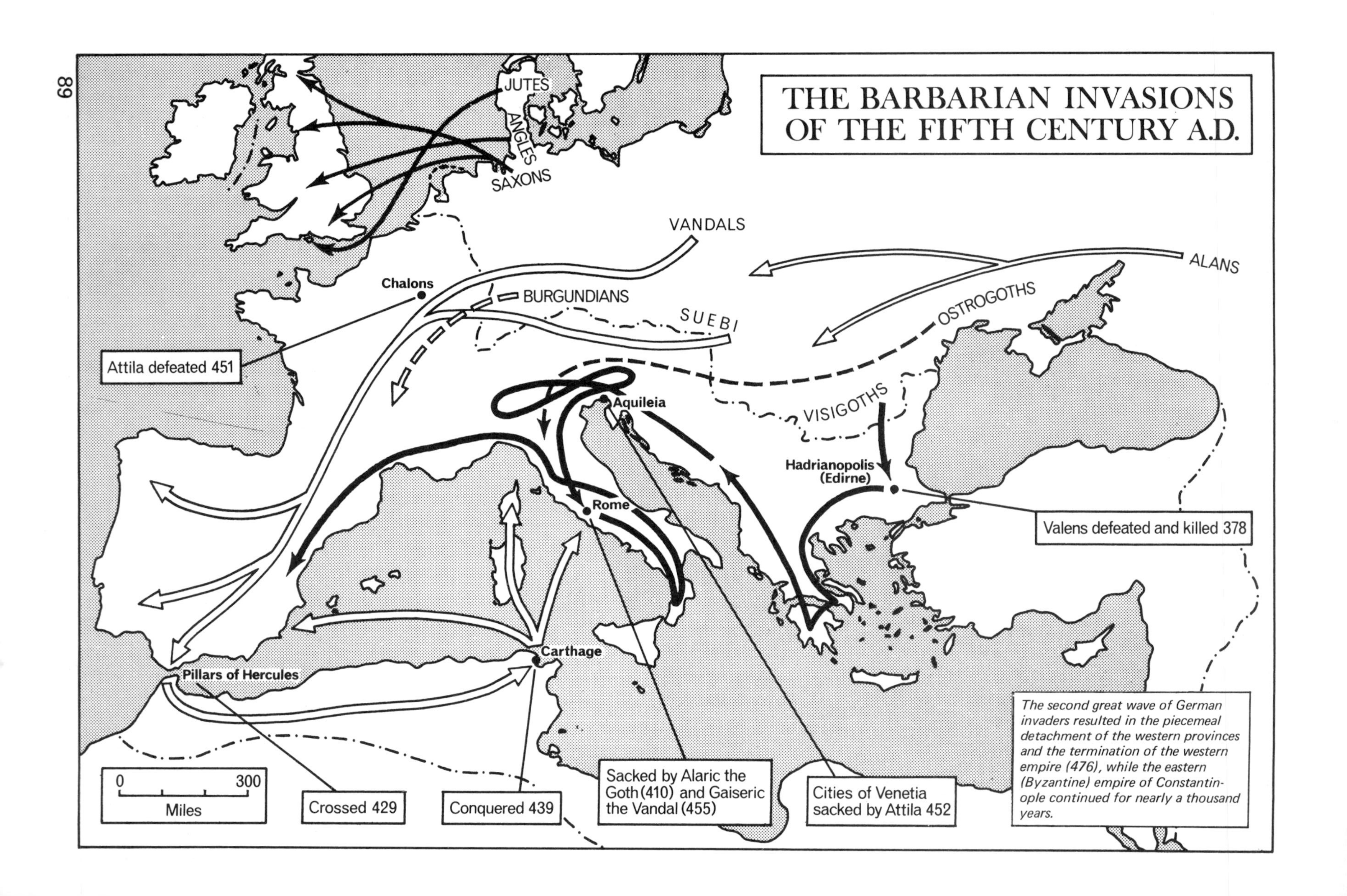
THE BARBARIAN INVASIONS OF THE FIFTH CENTURY A.D.
JUTES
ANGLES
SAXONS
VANDALS
ALANS
OSTROGOTHS
BURGUNDIANS
SUEBI
VISIGOTHS
Chalons
Attila defeated 451
Aquileia
Hadrianopolis (Edirne)
Valens defeated and killed 378
Rome
Carthage
Pillars of Hercules
0
300
Miles
Crossed 429
Conquered 439
Sacked by Alaric the Goth (410) and Gaiseric the Vandal (455)
Cities of Venetia sacked by Attila 452
The second great wave of German invaders resulted in the piecemeal detachment of the western provinces and the termination of the western empire (476), while the eastern (Byzantine) empire of Constantinople continued for nearly a thousand years.

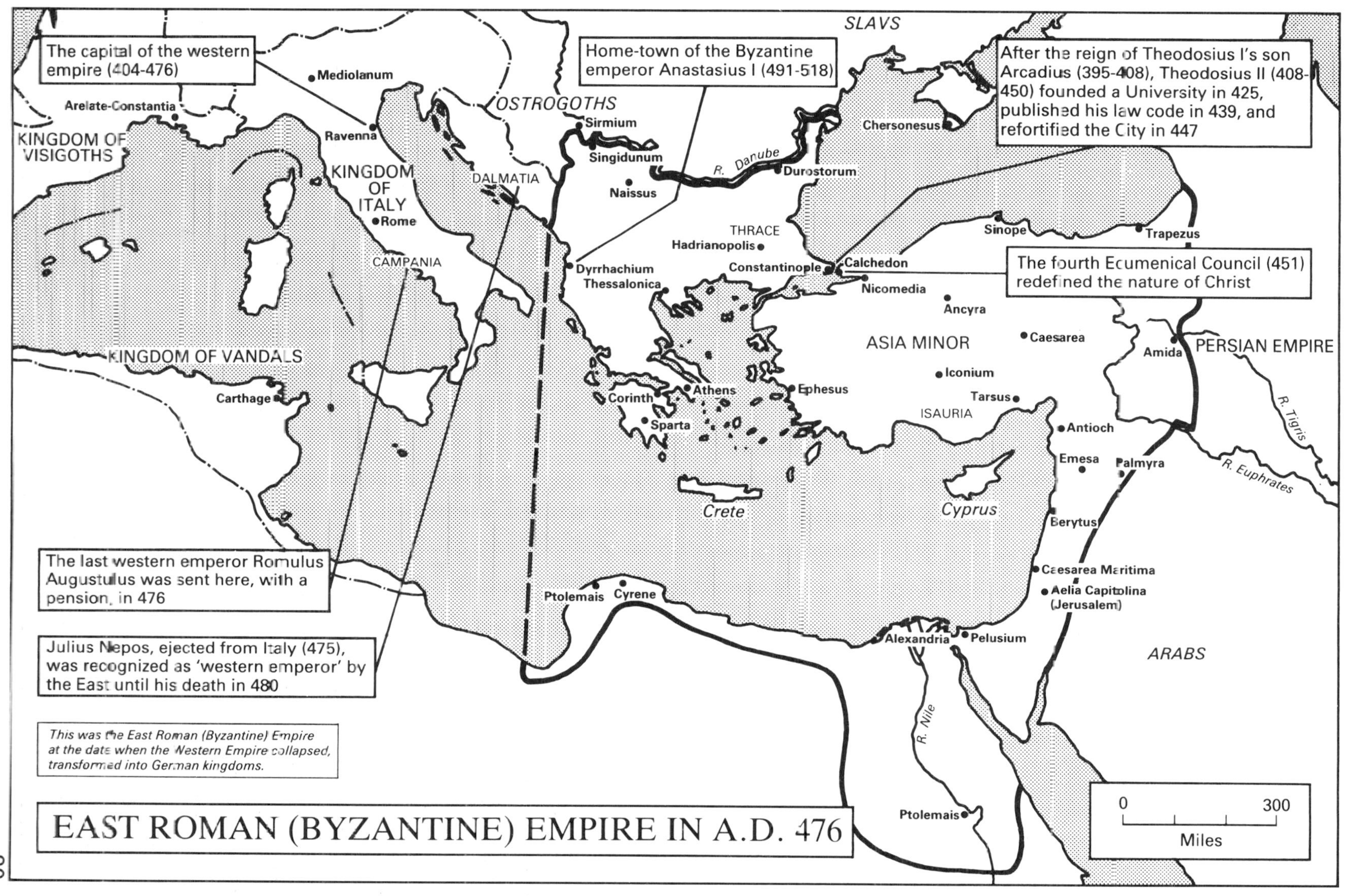

EAST ROMAN (BYZANTINE) EMPIRE IN A.D. 476

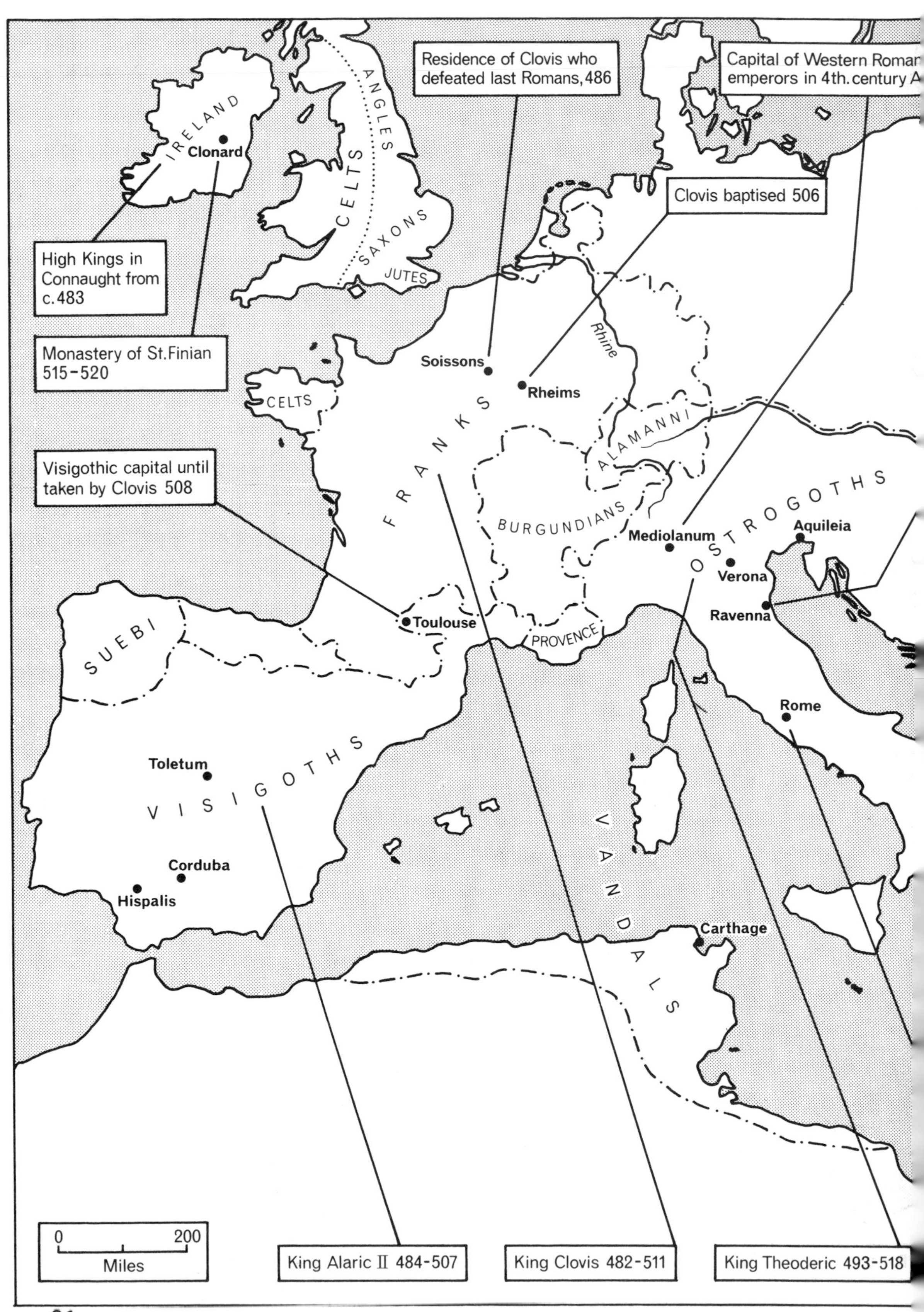

Residence of Clovis who defeated last Romans, 486
Capital of Western Roman emperors in 4th. century A
IRELAND
Clonard
CELTS
ANGLES
SAXONS
JUTES
Clovis baptised 506
High Kings in Connaught from c.483
Monastery of St.Finian 515-520
Soissons
Rheims
Rhine
CELTS
FRANKS
ALAMANNI
Visigothic capital until taken by Clovis 508
BURGUNDIANS
OSTROGOTHS
Mediolanum
Aquileia
Verona
Ravenna
Toulouse
PROVENCE
SUEBI
Rome
Toletum
VISIGOTHS
VANDALS
Corduba
Hispalis
Carthage
0
200
Miles
King Alaric II 484-507
King Clovis 482-511
King Theoderic 493-518

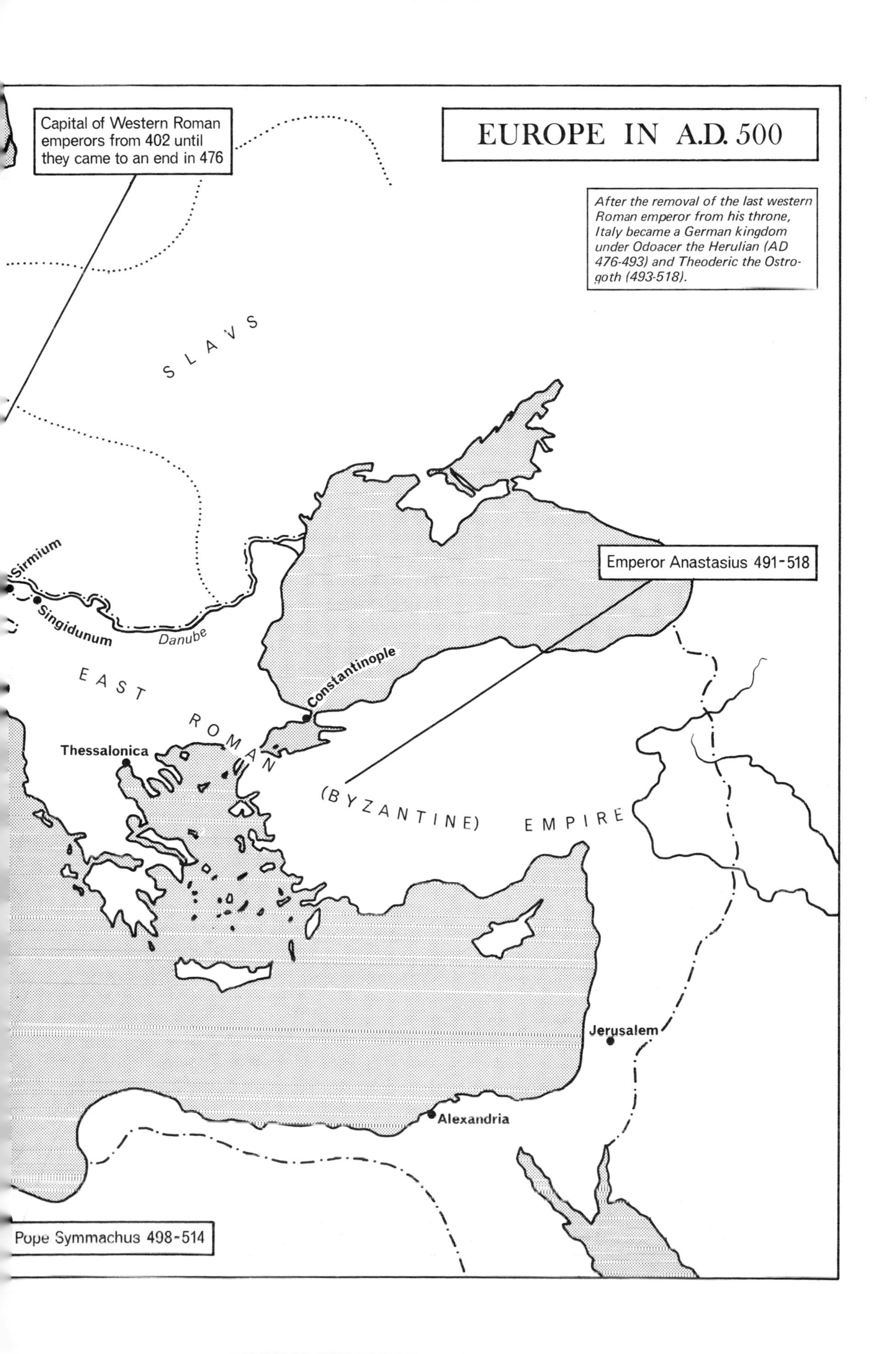
EUROPE IN A.D. 500
Capital of Western Roman emperors from 402 until they came to an end in 476
After the removal of the last western Roman emperor from his throne, Italy became a German kingdom under Odoacer the Herulian (AD 476-493) and Theoderic the Ostrogoth (493-518).
SLAVS
Emperor Anastasius 491-518
Sirmium
Singidunum
Danube
Constantinople
EAST ROMAN (BYZANTINE) EMPIRE
Thessalonica
Jerusalem
Alexandria
Pope Symmachus 498-514

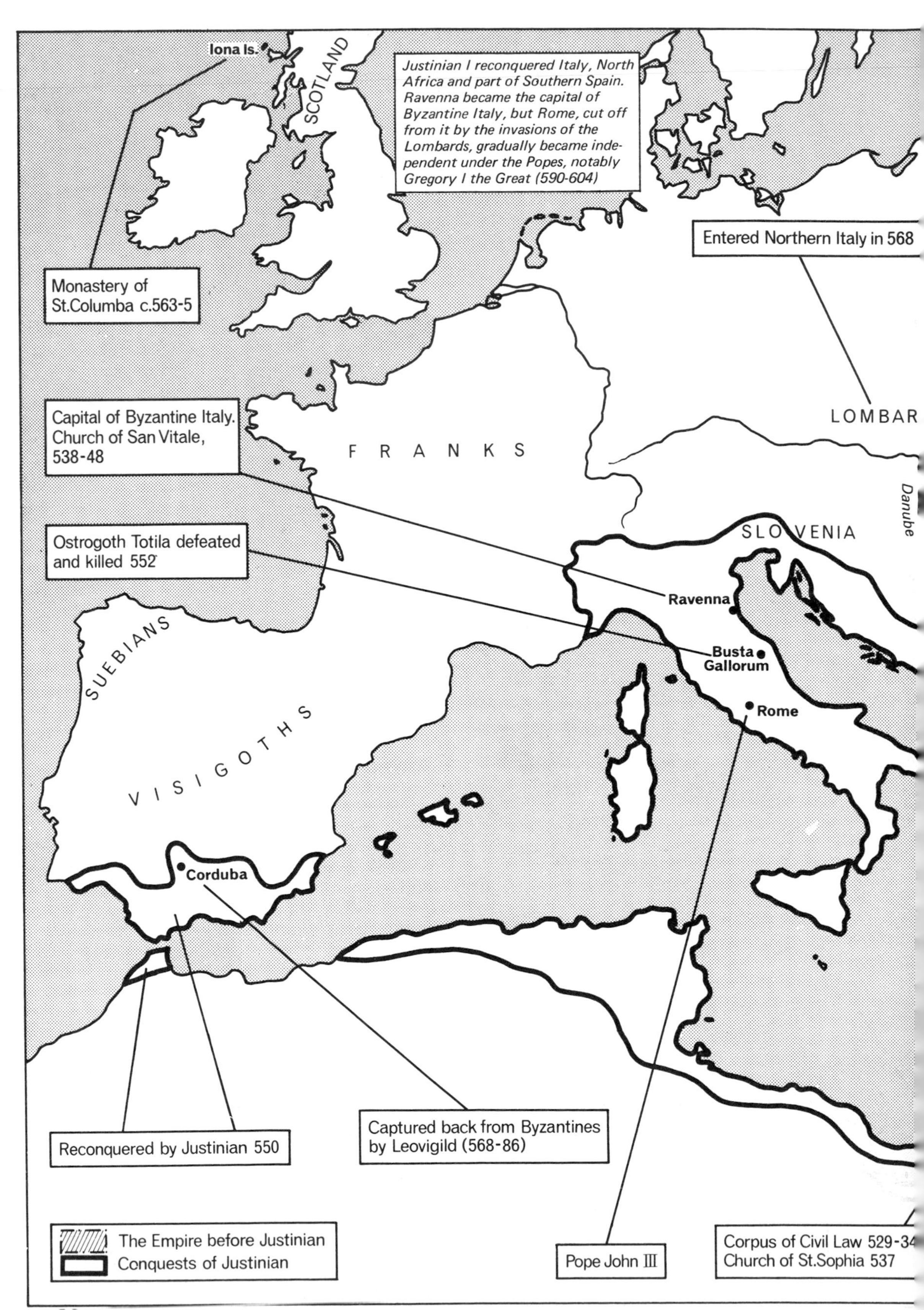
Iona Is.
SCOTLAND
Justinian I reconquered Italy, North Africa and part of Southern Spain. Ravenna became the capital of Byzantine Italy, but Rome, cut off from it by the invasions of the Lombards, gradually became independent under the Popes, notably Gregory I the Great (590-604)
Entered Northern Italy in 568
Monastery of St.Columba c.563-5
Capital of Byzantine Italy. Church of San Vitale, 538-48
FRANKS
LOMBAR
Danube
SLOVENIA
Ostrogoth Totila defeated and killed 552
Ravenna
Busta Gallorum
SUEBIANS
VISIGOTHS
Rome
Corduba
Reconquered by Justinian 550
Captured back from Byzantines by Leovigild (568-86)
The Empire before Justinian
Conquests of Justinian
Pope John III
Corpus of Civil Law 529-34
Church of St.Sophia 537

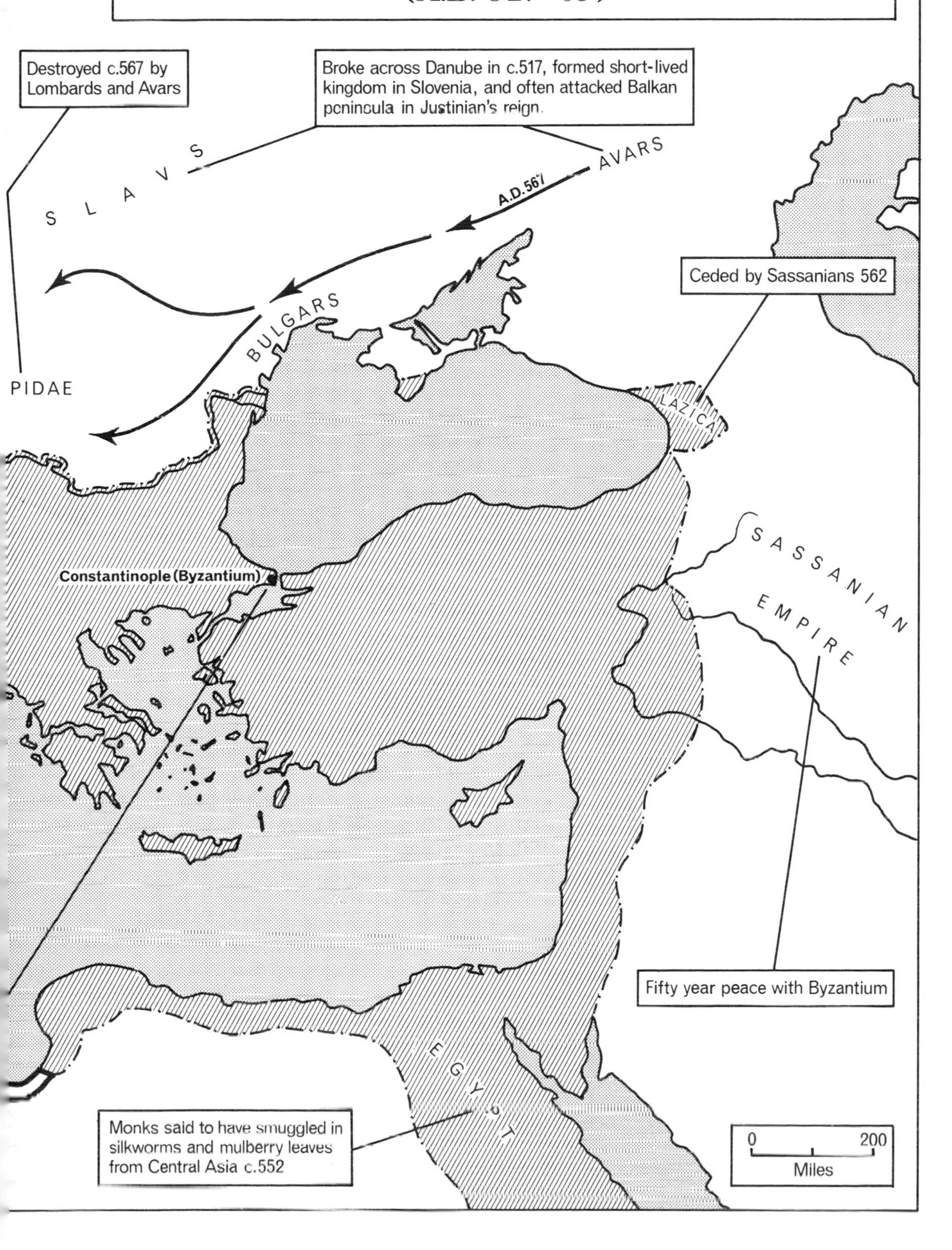

THE BYZANTINE EMPIRE OF JUSTINIAN I
(A.D. 527-65)
Destroyed c.567 by Lombards and Avars
Broke across Danube in c.517, formed short-lived kingdom in Slovenia, and often attacked Balkan peninsula in Justinian's reign.
SLAVS
AVARS
A.D.567
Ceded by Sassanians 562
BULGARS
PIDAE
LAZICA
SASSANIAN EMPIRE
Constantinople (Byzantium)
Fifty year peace with Byzantium
EGYPT
Monks said to have smuggled in silkworms and mulberry leaves from Central Asia c.552
0
200
Miles

Index of Place Names[1]

Modern names are given in brackets

[1] I have sometimes sacrificed consistency of spelling to convenience and tradition.